FOREST POLICY

FORESTRY SCIENCES

Books previously published:

Baas P, ed: New Perspectives in Wood Anatomy. 1982. ISBN 90-247-2526-7
Prins CFL, ed: Production, Marketing and Use of Finger-Jointed Sawnwood. 1982.
  ISBN 90-247-2569-0
Oldeman RAA, et al. eds: Tropical Utilization: Practice and Prospects. 1982.
  ISBN 90-247-2581-X
Den Ouden P and Boom BK, eds: Manual of Cultivated Conifers: Hardy in Cold and
  Warm-Temperate Zone. 1982. ISBN 90-247-2148-2 paperback; ISBN 90-247-2644-1
  hardbound
Bonga JM and Durzan DJ, eds: Tissue Culture in Forestry. 1982. ISBN 90-247-2660-3
Satoo T and Madgwick HAI: Forest Biomass. 1982. ISBN 90-247-2710-3
Van Nao T, ed: Forest Fire Prevention and Control. 1982. ISBN 90-247-3050-3
Douglas J: A Re-appraisal of Forestry Development in Developing Countries. 1983.
  ISBN 90-247-2830-4
Gordon JC and Wheeler CT, eds: Biological Nitrogen Fixation in Forest Ecosystems:
  Foundations and Applications. 1983. ISBN 90-247-2849-5
Duryea ML and Landis TD, eds: Forest Nursery Manual: Production of Bare Root
  Seedlings. 1983. ISBN 90-247-2914-9 paperback; ISBN 90-247-2913-0 hardbound

In preparation:

Németh MV: The Virus-Mycoplasma and Rickettsia Disease of Fruit Trees
Brown GN and Duryea ML, eds: Seed Physiology and Reforestation Success

Forest policy
A contribution to resource development

Edited by

F.C. Hummel

1984 **MARTINUS NIJHOFF/DR W. JUNK PUBLISHERS**
a member of the KLUWER ACADEMIC PUBLISHERS GROUP
THE HAGUE / BOSTON / LANCASTER

Distributors

for the United States and Canada: Kluwer Boston, Inc., 190 Old Derby Street, Hingham, MA 02043, USA

for all other countries: Kluwer Academic Publishers Group, Distribution Center, P.O.Box 322, 3300 AH Dordrecht, The Netherlands

Library of Congress Cataloging in Publication Data

Main entry under title:

Forest policy.

   (Forestry sciences ; v. 12)
   Includes bibliographical references.
   1. Forest policy.　I. Hummel, F. C.　II. Series.
SD561.F67　1984　　　333.75　　83–18242

ISBN-13: 978-94-009-6094-7　　　e-ISBN-13: 978-94-009-6092-3

DOI: 10.1007/978-94-009-6092-3

Table of contents

The key issues   xi
Acknowledgements   xiii
Introduction   xv

1 FORESTS AND FORESTRY IN NATIONAL LIFE   1
by Adriaan van Maaren

1.1   Introduction   1
1.1.1 Forest   1
1.1.2 Forestry   2
1.1.3 Forest policy   2
1.2   Past developments   4
1.3   The role of forests and Forestry   6
1.3.1 Forest functions   7
1.3.2 Forestry, the art of managing forests   9
1.4   Policy aspects of forest management   11
1.5   Forestry in relation to other policies   15
1.6   Co-ordination of policies   18
References   19

2 THE WORLD PERSPECTIVE   21
by Tim Peck

2.1   Introduction   21
2.2   The forest resource   22
2.2.1 Land use categories   26
2.2.2 Main types of forest   27
2.2.3 Standing volume   34

2.2.4 The outlook for the forest resource 35
2.2.5 Man-made forests 38
2.3   World trends in wood consumption and supply 40
2.3.1 Present consumption of forest products 40
2.3.2 The wood harvest 46
2.3.3 International trade in forest products 50
2.3.4 The growing interest in long-term outlook studies 51
2.3.5 Major long-term studies 57
2.3.6 The world outlook 58
2.4   International implications for national policies 62
Notes 65
References 66

3 THE PRODUCTION FUNCTIONS 67
by Otto Eckmüllner and András Madas

3.1   Forest products 67
3.1.1 Wood in the world economy 67
3.1.2 Factors influencing the consumption of industrial wood 69
3.1.3 The industrial wood products 71
3.1.4 Trends in the structure of wood consumption 76
3.1.5 Forest products other than wood 81
3.2   Forest production 83
3.2.1 Introduction 83
3.2.2 Past developments 84
3.2.3 Some policy aspects 84
3.2.4 The principle of sustained yield 86
3.2.5 The rotation period and its function 88
3.2.6 Natural forests and plantations 89
3.2.7 Permanent or interrupted biological production 91
3.2.8 How to increase wood production 91
3.3   Marketing 93
3.3.1 The market economies 93
3.3.2 The centrally planned economies 99
3.4   Forest policy and forest industry 105
3.4.1 Introduction 105
3.4.2 Forecasting supply and consumption 106
3.4.3 Forest industries – objectives and options 106
3.4.4 The relations between forestry and forest industry 108
References 126

4 THE SERVICE FUNCTIONS   127
by András Madas

4.1   Protection of the environment   127
4.1.1 The general situation   127
4.1.2 The water cycle   129
4.1.3 The atmosphere   133
4.1.4 The soil   139
4.1.5 Wildlife and landscape   141
4.2   Recreation   145
4.2.1 Introduction   145
4.2.2 General recreation   146
4.2.3 Specific recreational uses   148
4.3   Urban forestry   150
4.4   Evaluation and financing of service functions   152
4.4.1 Introduction   152
4.4.2 Methods of evaluation   153
4.4.3 Methods of financing   156
References   158

5 SOME SPECIAL TOPICS   161
by Otto Eckmüllner and Adriaan van Maaren

5.1   Forest protection   161
5.1.1 Introduction   161
5.1.2 Past developments   162
5.1.3 Causes of damage   163
      Forest clearance   163
      Over-exploitation   164
      Damage by game   165
      Damage by grazing   165
      Air pollution   166
      Forest pollution on the ground   167
      Fire   168
      Wind and snow   168
      Insects and fungi   169
5.1.4 General policy implications   170
5.2   The concept of biomass   171
5.2.1 Introduction   171
5.2.2 History   172
5.2.3 Policy aspects   173
5.3   Wood and energy   176

5.3.1 Introduction   176
5.3.2 Some facts and figures   177
5.3.3 Policy implications   179
5.4   Tropical moist forest   182
5.4.1 Introduction   182
5.4.2 History   184
5.4.3 Policy considerations   186
5.4.4 Policy options   188
5.5   Farm Forestry   190
5.5.1 Introduction   190
5.5.2 Types of farm forests   192
5.5.3 The dynamics in farm forestry   193
5.5.4 Forest policy and farm forestry   194
5.6   Agroforestry   196
5.6.1 Introduction   196
5.6.2 History   198
5.6.3 Policy aspects   200
5.7   Rural community development   203
5.7.1 Introduction   203
5.7.2 Developing countries   203
5.7.3 Developed countries   207
References   210

6 INSTITUTIONS AND ADMINISTRATION   213
by Fred Hummel

6.1   Forest ownership   213
6.1.1 State forests   213
6.1.2 Forests owned by individuals   214
6.1.3 Communal forests   216
6.1.4 Co-operative forests   217
6.1.5 Forests owned by industry   218
6.1.6 Forests owned by institutions   219
6.1.7 Policy considerations   220
6.2   Legislation   221
6.2.1 General considerations   222
6.2.2 Drafting and reviewing forest legislation   223
6.2.3 Specific legal issues (land, production, protection) 225
6.3   Taxation and incentives   227
6.3.1 Taxation   228
6.3.2 Incentives   230
6.4   Education and training   233

6.4.1 University level (forest officers)   233
6.4.2 Technical level (foresters)   235
6.4.3 Operational level (forest workers)   236
6.4.4 Refresher courses   238
6.4.5 Courses to obtain additional qualifications   238
6.4.6 Informing the public   239
6.5   Research   240
6.5.1 History   240
6.5.2 Scope   241
6.5.3 Organization   242
6.5.4 Policy priorities   244
6.5.5 International co-operation   247
6.6   The government forest services   248
6.6.1 Ministerial responsibility   248
6.6.2 Forest authority and state forest enterprise   249
6.6.3 Structures   250
6.6.4 Finance   252
6.7   Personnel   253
6.7.1 Professional, technical and administrative personnel   253
6.7.2 Forest workers   255
References   257

7 INTERNATIONAL ORGANIZATIONS AND CONFERENCES 259
by Eero Kalkkinen

7.1   The background   259
7.2   Worldwide governmental organizations   260
7.2.1 FAO   261
7.2.2 UNESCO   265
7.2.3 UNEP   266
7.2.4 UNCTAD   266
7.2.5 UNIDO   266
7.2.6 WMO   267
7.2.7 ILO   267
7.2.8 World bank   268
7.2.9 World food programme   269
7.3   Worldwide non-governmental organizations   269
7.3.1 IUFRO   269
7.3.2 IUCN   270
7.3.3 WWF   270
7.3.4 ICRAF   270
7.4   Regional and other groups   271

7.4.1 United Nations Regional Economic Commissions (ECE, ECA,
        ECLA, ESCAP, ECWA)   271
7.4.2 OECD   272
7.4.3 EEC   272
7.4.4 Regional development banks   273
7.4.5 Other regional groups   274
7.5   Problems and achievements   275
Annex: Addresses of International Organizations   279

8 POLICY FORMATION   283

by Fred Hummel and Adriaan van Maaren

8.1   The framework   283
8.1.1 General points   283
8.1.2 Responsibilities   285
8.1.3 Public involvement   286
8.1.4 Steps in policy formation   288
8.1.5 Implementation   291
8.2   Practice   292
8.2.1 The need for new policies   292
8.2.2 The direction of change – developing countries   292
8.2.3 The direction of change – developed countries   296
8.2.4 The process of policy formation   297
8.2.5 Concluding remarks   302
References   303
Annex: Objectives and Principles of Forestry Policy (Proposal by EEC
Commission)   305

The key issues

The book covers too wide a subject for a meaningful summary to be presented in a few pages, but there are a few basic ideas which give the book its orientation and which the prospective reader may wish to know at the outset.

1   In nearly all countries of the world *forest policies need to be strengthened and implemented more rigorously* because existing measures are insufficient to meet the rising demands made on forests for timber and other forest products, for the protection of the environment (soil and water conservation etc.) and for the provision of recreational opportunities. The possibilities for increasing production without prejudicing the environment are great. It is in the poorest countries, where hundreds of millions of people are short of wood to cook their food, that action is most urgent.

2   While continuing to pursue the traditional long term objectives of timber production and conservation of the environment, *forest policy should give much more emphasis* than in the past *to the immediate needs and wishes of people:* in the first place all who live in or near forests, forest workers, forest owners, but also all others who may benefit from the forest as consumers of forest products or holiday makers etc. That is why the book underlines topics such as wood and energy, agroforestry, rural community development, farm forestry and urban forestry.

3   *There must be a close co-ordination between forest policy and other policies which influence or are influenced by forestry;* among the more important are the policies for agriculture, environment, regional development, industry, tourism, and employment.

4   *Forest policy must be treated as a coherent whole.* Only in this way can a sensible balance be achieved between the production, environmental and recreational objectives which should as far as possible be pursued by *multipurpose management.* A coherent overall forest policy is also essential if damaging distor-

tions and inconsistencies are to be avoided which can arise from the links with other policites. For example, policies for industry, tourism and the environment often seek to pull forestry in different directions.

5 *National forest policies should recognize the growing importance of forestry's international dimension,* especially the impact of imports and exports of forest products and certain environmental aspects. For example: watershed management in one country affects the water regime of its neighbour; the survival of endangered species of animals and plants is of world concern and forests are one of their main habitats. In this respect tropical moist forest is particularly relevant; it is estimated to contain some three million species of which only 10-15% have been identified and named. Development aid is another major aspect of forestry's international dimension, but also the advantages of co-operation between countries at a similar stage of development must not be ignored. Both national governments and international organizations have an important role to play in these matters.

6 *Those responsible for forest policy should do what they can to protect forests against avoidable damage and destruction;* where the responsibility for the necessary action rests with others, foresters should at least exert the maximum influence they can. Protection against atmospheric pollution is a topical example.

7 *Forest administrations and individual forest officers should become more outward looking. Without becoming less concerned with trees, they must become more concerned with people.* It is only in this way that forestry can play its proper part in national life and on the world stage and thus effectively serve mankind.

Acknowledgements

We authors have benefitted greatly from the generous responses to our numerous requests for information, advice and photographs. Indeed some parts of the book could not have been completed without this valuable help.

In the first place we thank Mr M. Flores Rodas, Assistant Director General in charge of FAO's Forestry Department and his colleagues. Messrs M.K. Muthoo, J.E.M. Arnold and P.R. Wardle kindly commented on some of the drafts and supplied information, while Mr T.M. Pasca let us select photographs from the archives of UNASYLVA of which he was editor at the time.

Also the following were good enough to offer observations on various drafts: Mr. C.J. Lancaster (UNDP); Mr J.S. Spears (World Bank); Dr. K. Oedekoven (former head of forest service, Federal Republic of Germany); Professors J. Speer and R. Zundel (Federal Republic of Germany); Professor R. Morandini (Italy); Professor J. Lammi (USA); Dr. V. Holopainen (Finland); Mr. J. Douglas (Australia); Messrs P. Adlard and D.R. Johnston (UK).

We are very grateful too for the photographs put at our disposal by: Baron and Baroness von Aretin (Federal Republic or Germany); Dr. J. Evans, Miss. Julia and Mr. Antony Hummel; British Columbia Forest Service; Forestry Commission (UK); US Forest Service; Western Softwood Ltd (UK).

Conversations and correspondence with dr. ir. J.H.A. Boerboom, drs. A.M. Filius, ir. C.P. Veer and ir. K.F. Wiersum of the Forestry Departments at Hinkeloord Wageningen as well as with many others who cannot all be named individually have also added greatly to our knowledge and have influenced our views; to these too we extend our most cordial thanks.

Last but not least we are much indebted to all who contributed to the processing of the manuscripts, especially to Mrs Anne Case who typed the final drafts and helped in many other ways.

Introduction

This book places forest policy into the broader context of land use and the management and conservation of the world's renewable natural resources in a changing world. World trends and national forest policy options are considered against the background of the increasing pressures to which the forests are being subjected by rising populations and man's quest for a better life – sometimes with too little thought for the future. The production of timber and other forest products, as well as the environmental and other service functions of forests, are discussed and so are the process of policy formation and implementation and other administrative and institutional aspects of forest policy.

The book is intended primarily for administrators and politicians with an interest in resource management and conservation, as well as for foresters and students of forestry who want to look beyond the trees and beyond their own country; the major issues with which the book is concerned will, it is hoped, also provide food for thought to forest industrialists, conservationists, sociologists and indeed to all with a concern for the future of our planet as a fit place for our children and grandchildren to live in.

Few, if any, elections have been won or lost, few political careers made or marred by issues of forest policy; one reason is that while perhaps recognized as important, these issues are rarely regarded as urgent. The benefits of wise action and the harm done by wrong action – or more frequently inaction – may not become apparent for many years. There is thus always a temptation to accord to forestry measures a relatively low priority, even when they are uncontroversial. Perhaps the cause of forestry has also not been helped by premature or exaggerated forecasts of impending world shortages of timber and environmental catastrophe. This book attempts to demonstrate that the position is very serious indeed even without exaggeration.

In its overall view of the major issues of forest policy, the book complements and draws on the mass of literature which deals with the forest policies of particular countries or with particular aspects of forestry over a wider area. As far as the authors are aware, this is the first published attempt at such a broad approach since FAO's 'Forests Policy, Law and Administration' appeared in 1950 (FAO Forestry and Forest Products Studies No 2). The basic functions of forests are still the same as they were then namely

- to produce timber, firewood and a large variety of other products;
- to provide services and more especially to contribute to the conservation of the environment, including soil, water, landscape, animals, plants and atmosphere.

But there have been major developments in the problems and policies which relate to these functions. To mention some of the more important ones: The consumption of industrial wood in the world has more than doubled and is still rising; a shortage of firewood brought about by forest destruction and a population explosion has created a poor man's energy crisis in many developing countries; the destruction of the remaining tropical moist forests is continuing to gather momentum with serious economic, social and environmental consequences for the countries concerned, and possibly also for the world's atmosphere; the shift in emphasis towards what has become known as forestry for the people, has encouraged major developments in forestry for rural community development, urban forestry, agroforestry and the recreational use of forests; the more intensive use of land outside the forest for agriculture and urban purposes, as well as the more intensive management of the forests themselves has threatened the habitats of numerous species of animals and plants and endangered their survival; above all, the accelerating rate at which the world's non-renewable resources are being exploited, has added to the urgency of developing the full potential of the renewable resources, of which forests are amoung the most important. At the same time, the very existence of forests appears to be threatened in some of the more industrialized and densely populated regions of the world by the sudden and large scale dieback of trees from causes as yet not clearly identified but probably connected in some way with the rising pollution of the atmosphere.

At the risk of some slight repetition each of the eight chapters of the book has been made as self contained as possible in order to minimize the need for cross referencing. *Chapter 1* describes the role of forests and forestry in national life; it sets out the basic forest functions, summarizes past developments of forest policy and discusses the links with other national policies. *Chapter 2* gives the international perspective; it contains the basic statistics and other information on the world's forest resources, timber consumption, supplies and trade, together with forecasts of trends and their implications for national policy options. *Chapter 3* is devoted to the productive functions of forests; it starts with the consumption trends for various groups of forest products and goes on to describe the various ways in which production may be varied in quantity, quality and timing; marketing is dealt with in two sub-sections, one for the market economies and the other for the centrally planned economies; finally there is a section on the implications of forest policy for forest industries. *Chapter 4* outlines the role of forests in the conservation of nature and the provision of recreational opportunities; it points to the possible conflicts that may arise between these functions and the productive function and suggests policy options for minimizing these conflicts by the concept of multipurpose management or in other ways. The chapter also discusses the role of forests in global environmental problems, such as the oxygen and carbon dioxide cycle of the atmosphere. *Chapter 5* starts with a section on forest protection and then highlights a number of topical issues, which it was thought best to single out rather than to incorporate in the general body of the text. They are: forest protection, biomass, wood and energy, tropical moist forest, farm forests, agroforestry and forestry for rural community development.

While chapters 1 to 5 are concerned mainly with technical, social and economic aspects of forest policy, chapters 6 to 8 are devoted to administrative and institutional aspects and to the process of policy formation. *Chapter 6* covers forest ownership, legislation, taxation and incentives, education and training, research, organization of forest services and personnel. *Chapter 7* outlines the roles of international organizations and conferences and their influence on national forest policies. The final *Chapter 8* opens with the general principles of policy formation and concludes with the actual methods by which governments review and update forest policies.

The scope of this book, the balance between the topics covered and the form of presentation require some explanation. First, the title: *Forest* refers to associations of plants and animals in which trees are dominant, including land with partial tree cover or scrub vegetation, such as tropical savanna or mediterranean maquis. *Policy,* is used in the sense of the following definition given in Webster's 3rd New International Dictionary: 'a definite course or method of action selected (as by a government, institution, group of individual) from among alternatives and in the light of given conditions to guide and usually determine present and future decisions.' As most important policy decisions are taken at national level, the book refers mainly, although not exclusively to national policies. Activities associated with the management of forests are referred to as *forestry*. Because forest policy influences and is influenced by other policies, e.g. those concerning agriculture, forest industries, regional development and environment, the implications of these links could not be ignored; this applies especially to topics such as agroforestry and urban forestry where the need for co-ordination with other policies is most pronounced. The balance between topics has, if anything, been tipped in favour of new developments but, it is hoped, without distracting attention from fundamentals of policy which have withstood the test of time. A balance had also to be struck between discussing the theoretical foundations of forest policy and the options with which the policy makers of today are confronted in practice. The main emphasis has been placed on the latter. The theoretical foundations are rooted in several branches of science dealing with the manifold connections between nature, society, economy on the one hand and forestry on the other hand. The object of studying these scientific foundations is to define the general situation, the multiple functions of forests, the measures necessary to meet the demands made on forests and the manifold legal, planning, organizing, and other actions necessary in this context. The policy options described are based in part on the above foundations, and in part on what is actually happening or should urgently be done in different parts of the world.

The examples given in the text from various parts of the world are by way of illustration only, and also the references to literature are highly selective. To have attempted a more complete coverage would have defeated the aim to highlight the essentials which are only too easily buried in detail. In the circumstances, we felt that no useful purpose would be served by an index; but in order to facilitate references to particular topics, the table of contents is detailed.

However objective authors may try to be, their personal backgrounds and experiences must colour their views, Readers may, therefore, wish to know who we are. In alphabetical order:

O. Eckmüllner, Dipl. Ing. for., Dr. After studying forestry in Vienna, he began his forestry career as a member of the Forestry Department of the Chamber for Agriculture and Forestry in Styria, where he worked later on – after an interruption of six years (World War II and prisoner of war in the USSR) – as forestry director of the said department until 1962. From 1954 he was at the same time 'Lehrbeauftragter' (lecturer) for farmforestry in Vienna. In 1962 he was appointed professor of Forest Policy at the Agrarian University of Vienna. He retained this professorship while from 1967 to 1973 he was also head of the Austrian Forest Service in the Ministry of Agriculture and Forestry. For 25 years he was Austrian delegate at the European Forestry Commission of FAO in Rome and of the Timber Committee of the Economic Commission for Europe in Geneva. For four years he was chairman of the European Forestry Commission.

F.C. Hummel, M.A., D. Phil (Oxon), Dr. h.c. (Munich): After studying forestry at Oxford, he began his career in 1938 as a district forest officer in Uganda. In 1946 he joined the Forestry Commission in Britain where he held the posts of mensuration officer and later chief of the planning branch. From 1961-1966 he worked for FAO as the Codirector of the Mexican National Forest Inventory. He then returned to Britain where he was promoted to the post of Commissioner for Harvesting and Marketing in 1968. In 1973 he joined the EEC Commission in Brussels as Head of the Forestry and Environment Division, a post which he held until his retirement in 1980. Since then he has undertaken consultant assignments in Europe, Nigeria, Bangladesh and Mexico. In the course of his career he has contributed articles on forest policy, planning and mensuration to forestry journals in several countries of Europe and elsewhere and he has also undertaken a number of lecture tours and consultancies.

E.F.I. Kalkkinen, B.Com. After studying timber trade in London (1938/39) he served in the Finnish Army (1939/44). On completion of his university studies in

Finland he worked in the timber trade in Belgium (1946/48). In 1948 he joined the FAO/ECE Timber Division in Geneva as forest products market analyst and economist. In 1958 he was appointed regional forest economist for the Latin American region of FAO, and in 1960 regional forestry officer for the same region. In 1963 he came back to Geneva as director of the FAO/ECE timber Division, and was appointed in 1977 director of the FAO/ECE Agriculture and Timber Division following the merger of FAO's two joint divisions with the ECE. He retired in 1982. In the course of his career, Kalkinen participated in the work of most·international organizations and bodies dealing with forestry, forest products and forest economics. He has given many lectures and conferences in different countries on subjects related with his work. He also participated in the preparation of the timber trends studies which were published during his years of service with FAO and ECE.

András Madas, Dr. Acad. Scis. Hungary: After studying forestry at the University of Sopron, Hungary, 1935-40, he began his career as forest engineer in construction at the State Forestry Directorate in the Carpathians. In 1949 he joined the State Planning Committee in Budapest as Chief of the Forestry Sector; in 1960 he was appointed director of the Agriculture, Food, Forestry and Water Management Division. In 1972 he became Deputy Minister in the Hungarian Ministry of Agriculture and Food as Chief of the International Affairs of the Ministry and the Forestry Branch. For 22 years he was the head of the Hungarian delegations at the Timber Committee of the Economic Commission for Europe in Geneva of which he was elected vice-Chairman for 1970-73 and Chairman for 1974-76. At the VII World Forestry Congress in Buenos Aires, 1972, he was Chairman of 'Commission VI: The Economists, Administrators and Planners'. Between 1972-75 he was the head of the Hungarian delegation at the Council of Forestry of the COMECON. His main publications are: World Consumption of Wood, 1974 (in English); Forestry Policy, 1978, (in Hungarian); LES BOIS: Nouvelles perspective pour les pays en voie de development. Revue Tiers-Monde, 1976 (Study in French, Paris); Forest Sector Planning in Centrally Planned Economies, 1971. For many years he has lectured at the University of Sopron where he ranks as a university professor.

T.J. Peck was educated at Sherborne School and the University of Oxford (United Kingdom). On graduating with a M.A. (Forestry) in 1956, he worked with a firm of forest management consultants in central England until 1959. From that year until the present, he has been employed by the UN Food and Agriculture Organization with the FAO/ECE Timber Division (later to become the FAO/ECE Agriculture and Timber Division) in Geneva, Switzerland. During the 1960's and early 1970's, his work was mainly concerned with forest products market analysis. During sabbatical leave in 1973/74 he studied at the University of British Columbia, Vancouver, Canada, Specializing in environmental aspects of forestry and forest industries and gaining the degree of Master of Forestry. On returning to Geneva, he acted as editor and principal author of the third in the series of FAO/ECE long-term studies on Europe, published in 1976 as 'European Timber Trends and Prospects, 1950 to 2000'. In 1978, he was appointed Chief of the Timber Section of the FAO/ECE Agriculture and Timber Division, with overall responsibility for the programmes of the FAO European Forestry Commission, the ECE Timber Committee and their joint subaidiary bodies. His work has taken him to practically all countries in Europe, as well as to the USSR, Canada, the USA, Mexico, Australia, New Zealand, Singapore and India.

A. van Maaren, ir. (Dutch equivalent for Masters' degree) was born in 1928. After studying forestry in Wageningen, with emphasis on the tropics he began his career with two years forest production research. From 1954 until 1972 he acted as a district and provincial forest officer in The Netherlands. He then joined the Directorate Board of the State Forest Administration, in charge of all Forestry in The Netherlands and the related international affairs. He took part in the 7th and 8th World forestry Congresses in Buenos Aires and Jakarta as well as in congresses and symposia in several European countries. In 1977 he was appointed Head of the Forest Management Department at Wageningen University. His lectures and research are in the field of forest management and policy. Since 1978, he represents The Netherlands in its co-operation project with the Forestry Faculty of the Gadjah mada University in Yogyakarta, Indonesia, with special emphasis on forestry and tree planting in the rural upland areas and on nature conservation on Java. Other departmental activities link his work to the rainforest exploitation on Kalimantan (Borneo) and agroforestry in the Machakos region (Kenya).

We authors, it will be seen, come from different parts of Europe and, as a group, we combine experience of national and international forest administration, education, research and practical forest management. Two of us have worked in Africa, Latin America and South East Asia, but we recognize that we see the world with European eyes and in a European perspective. We also recognize that what we have set out to do is very ambitious; and we hope that, in spite of whatever shortcomings the book may have, it wil assist its readers to promote the evolution of policies for forestry which respect the principles of sound resource management and correspond to the needs of a changing world.

1 Forests and forestry in national life

Adriaan van Maaren

1.1 Introduction

Forests are the most developed and widespread vegetation cover on our planet. Furthermore, every day almost every individual out of the myriads of people in the world is, wittingly or not, using or consuming some product or service from these forests.

The role of forests and forest management in every-day-life is thus obvious. However, questions will arise and discussions begin of the shrinking forest area and the lack or even the failures of forestry are taken into consideration. How to safeguard the availability of the forest products and the other blessings of the forest cover?

1.1.1 Forest

The word *forest* has many meanings but nowadays usually refers to *an association of plants and animals in which trees are dominant*. This is also the meaning of forest in this book. Forests are *'closed'* if the trees cover most of the ground and *'open'* if they are more scattered; they are *'in use'* if exploited or managed and *'not in use'* if *virgin* or *abandoned*. Small areas of forest are commonly referred to as *woodlands*. Most forests that are *managed* are closed forests but there are exceptions, for example substantial areas of tropical savanna are also under systematic forest management.

The designation by a government of land as forest is an act of policy, usually with legal implications under forest legislation. Outside these forests, however, are many smaller woodlands, hedges, rows and clumps of trees or even single trees which often play an important part in national life and should therefore also have the full attention of forest policy although they may not be the direct responsibility of the Forest Authority, but of some other ministry. That is why this book will deal not only with forests but with all land that can be considered to contribute substantially to the type of produce or the services that forests can supply.

1.1.2 Forestry

According to the definition adopted by the Society of American Foresters in 1967, the term forestry embraces:

> 'the management and use for human benefit of the natural resources that occur on and in association with forest lands'.

This definition has become too restrictive, because forestry is no longer confined to 'forest lands'. The small woodlands, windbreaks, hedgerows and even single trees on other land are also relevant. *Agroforestry* has in fact become an important discipline. Furthermore, forestry plays a vital part in the reclamation of devastated and eroded areas as well as in the afforestation of other bare land for the production of timber. The role of trees in the regreening of urban and industrial areas and the sides of motorways has given rise to *urban forestry*. For these reasons the definition of the term forestry has been widened for the purpose of this book to read as follows:

> *'Forestry is the management and use for human benefit of the natural resources that occur on or in association with forest lands and other lands managed wholly or in part for similar purposes.'*

1.1.3 Forest policy

Why should a nation develop its own 'policy', which will steer the availability of forests as well as the results of forestry into the whole of national life? First, what do we mean by policy, a word that means many things to many people? In this book policy denotes:

> 'a definite course or method of action (as by government, institutions, groups or individual) from among alternatives and in the light of given conditions to guide and usually determine future decisions.' (Webster's 3rd New International Dictionary)

Indeed, policies are adopted and followed, not only by governments but also in several places in society, eg in industry, by land owners and in villages. In chapters 6 and 8 special attention will be paid to this issue as far as forests and forestry are concerned. This book will however, concentrate mainly on policies decided upon by the national government.

The emphasis will be on the need for the formation and the implementation of a suitable forest policy, based on the relevant facts and on the views of all who are directly concerned as well as of the general public. A first step in the formation of policy is the identification of options and the formulation of proposals. The formation of forest policy is a continuous process designed to maintain the balance between the forest resource as the potential supplier on the one hand and the various components of society as the consumers on the other hand.

In the diagram on page 3 forest policy is pictured in the centre of a web of interrelations. The arrows symbolize the flow of information, instructions or decisions (3-11) as well as the results in products or services (1 and 2). The diagram is meant to give an overall view of the complicated character and the coherence of

Diagram: Framework of interrelations around forest policy. (according to M. de Coulon, 1977).

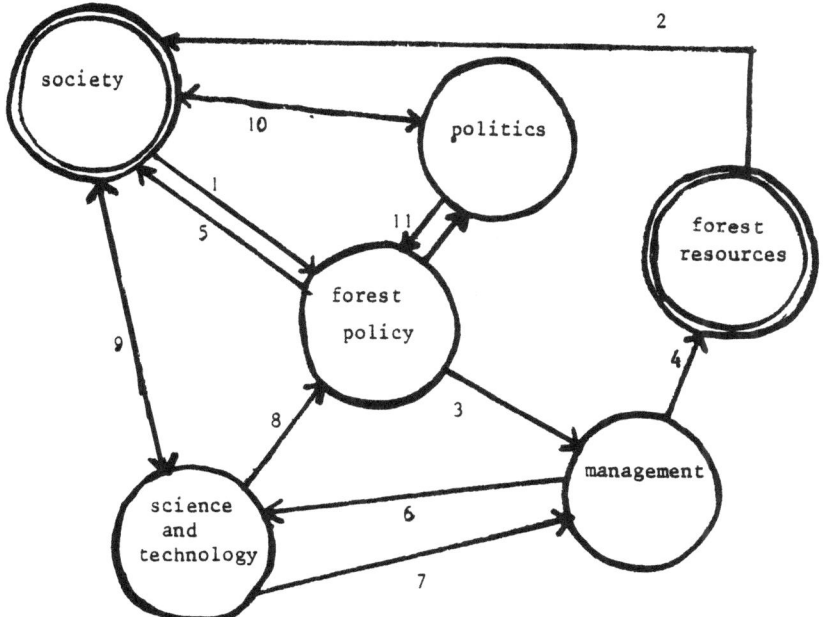

1 needs for forest products and services (consumption and expectations)
2 output of the forest resource (functions)
3 instructions for management (e.g. objectives or goals)
4 treatment and use of the forest resource
5 policy influencing society demands
6 problems to solve by science and technology
7 expertise
8 scientific basis for policy
9 interaction between society and science (research/education)
10 party-politics
11 political decision-making

the policy processes and all information that should be available. Most of the several parts of the webb will receive special attention in one or more chapters of this book. One thing, however, cannot be stressed too strongly:

> *forest policy has to balance in one way or another the relation of society with its forest resource.*

Even in the most simple situation, for instance a tribe living in a forest, the link between the people and the forest resource will run via some policy ideas or customs, activities and management.

1.2 Past developments

A brief look at history can help us to understand matters as they are today. On most land, forest is the natural vegetation cover. Consequently, as long as there are plenty of forests in relation to the sparse population and its needs, it matters little what kind of government policy exists. In such a situation, forests will suffer little or not at all from use by mankind.

During long periods in most regions of the world it was adequate for the ruling power to introduce a few regulations to safeguard its hunting rights. Subsequently, however, especially where the forest area became too small, the local ruler or landlord felt obliged to close the remaining woodlands to unlimited use by the local population. In the early days the 'closure' was generally restricted to the prohibition of pulling up trees; mostly no form of management was introduced. As a result, outside these forests, the woodland was more or less mined for household and industrial (charcoal) purposes, used for free-grazing animals like goats, or converted to pasture, arable land, villages, and so on. These developments caused shortages of wood for fuel and construction purposes. Dry areas became infertile or were even turned into deserts; in mountainous areas floods, landslides and other forms of erosion occurred, threatening the rural areas.

From that time, the need for the regulated management of forests became obvious and in that way 'forestry' was born. Alas, the ruins of former highly developed cultures are witnesses of how wrong things can go, for want of an adequate forestry policy and management.

The several factors which influence the stability or the decline of the forest vegetation are evident from the history of the Mediterranean area over the past three thousand years. The rise and decline of civilizations, the alternation of periods of stable government with longlasting instable war periods, growing populations and unadapted land use like the free-grazing goats, all elements can be traced, thanks to numerous historians throughout the period. In Central and Western Europe, since the 18th century the management and the use of forests has been developed on a sustained yield basis for the benefit of mankind. Canada symbolizes one of the most recent changes from laissez-faire forest exploitation to a policy of forest management.

However, the rising populations prevented the reforestation of all former woodlands. Vast areas remained in use for agriculture, grazing and the like. Especially in those cases where nearly all land was suited for agriculture in one form or another, only remnants of the former natural vegetation cover were left; this happened, for example, in Ireland and the Netherlands where three factors were mainly responsible:

- agriculture provides higher and more direct profits for the occupier of the land than forestry (these profits in agriculture are, however, due in part to heavy subsidies!);
- timber and other forest products could easily be imported from elsewhere;
- no negative externalities such as erosion hampers the agricultural land use in these two countries.

Generally, however, the necessity for a local or more widespread policy concerning forests and forestry was developing. In the early days, as said already, the preservation of the local ruler's rights, e.g. to safeguard wildlife for hunting and the availability of wood for housing purposes, may have seemed adequate. Where there was danger of floods, landslides, inland dunes and the like, however, more regulations were introduced in the common interest, e.g. the limitation or prohibition of grazing by domesticated animals.

Although the environmental aspects were not ignored, generally the pace at which forest policy developed and regulations introduced depended largely on timber shortages and periods of stable government. The late 17th century regulations and instructions in France, in favour of growing oak timber of large dimensions for naval purposes (Colbert) are well known. The still existing forests of mature oak (la chene royale) are showing the results of these former policy actions. The inland dune formations, such as were caused by overexploitation of woodlands in Northern Germany and the Netherlands, became a threat for the rural populations in the 17th century.

It was well into the 19th century before any serious action was taken on these environmental issues. The reason may be that the national government was little interested in such environmental disasters of only local importance. In contrast, as soon as it seemed profitable to afforest such areas, the inland dune fixation was effected within some decades.

A recent forest policy development specifically for strategic reasons occurred in the United Kingdom after World War I. Until that time, the British government relied on the possibility of easy timber imports. During the war, however, the U-boat blockade highlighted the danger of a too heavy dependence on such imports. After that war, the strategic aspect of forest policy and home-grown timber became a strong pillar of the UK forest policy for several decades.

Until World War II the forests in the Vosges near the frontier between France and Germany were another example of the strategic significance of some forests in modern times.

Another historical point needs to be stressed. As long as forests were plentiful and underused, local regulations by a landlord or a farming community were sufficient, as the problems that arose were mostly related to user rights. Along with the development of nation states, however, the national governments started to safeguard national interests concerning the supply of raw material, over-ruling the more local interests. Many forests were declared to be land of the king, royal forests; this led to centrally regulated policies as mentioned already in the case of France. In several countries of Western Europe this badly needed central policy evolution was retarded by political upheavals. Unfortunately, nowadays this kind of disruption impedes forest developments in many other parts of the world. The benefits for forestry of a stable national regime is enhanced where the more or less authoritarian approach that was customary in former times is replaced by more democratic procedures including the delegation of power to local authorities to deal with local questions. Indeed, the detailed nuances of forest policy should be handled locally, within the broad framework of a national or even supra national forest policy. The importance of such a frame-work is to secure the stability of the policy, a must for the maintenance of the optimal functioning of forests.

In modern times, national government policy for forests is based on the various roles of the forest and of the forestry, but a prerequisite for the conservation of the forest area and the introduction of management is the clear definition of the rights of ownership and use.

Forests have a *production function*: i.e. the supply of timber and other forest produce; and they have *service functions* such as the prevention of erosion.

The next section explains how the production and service functions can be manipulated so as to best serve the needs of growing populations and diversifying societies. The lesson of history that, the more diversified the need for special products and services and the bigger the required quantities, the more urgent is a policy to regulate the continuous provision of these products and services.

1.3 The role of forests and forestry

Generally, the role of forests and forestry in any country or region will be settled by two poles, the forest resource, i.e. the potential supply on the one hand and the need by society for forest products and services, i.e. the demand on the other hand. Neither supply nor demand are static. The forest resource can be managed for the purpose of optimal supply, but with time, society changes its use of products and services, in quality as well as in quantity. The local forest resource limits the possibilities of supply. When demand exceeds supply in a country or region, the difference must be imported if the demand is to be satisfied. Forest resources vary enormously from country to country because of climate, site and human interference which in most parts of the world has modified or destroyed natural forest ecosystems and reduced the original forest area.

The cultural influence of the forest has been recognized throughout the ages, and poets, painters, and philosophers have derived inspiration from it. The forest has left a deep impression on the minds of men from the earliest times, and is associated with religious beliefs among the people of many lands; this finds expression in the preservation of sacred groves and in the veneration of certain kinds of trees, while the folklore of many primitive tribes is bound up with the forests and the spirits which inhabit them.

Forests are subject to all kinds of activities initiated by society or parts of it. On the other hand, society is neither static nor uniform in constitution and must be considered as an amalgam of social systems. Every social system is, in a sense, a group of human beings interacting and sharing some common interests. Not only a family, a household, a school, a religious community, a village can be considered as forming such groups, but also a political lobby, the board of a company and even, subject to some qualifications, a nation. Each type of group, however, will make its own demand on the forest. Indeed, in one way or another, many people will have an interest in forestry including those concerned with industry, trade, tourism, and so on.

1.3.1 Forest functions

The relationship between the forest resource and social demand towards that resource is usually defined by so-called *forest functions*. These functions are in fact human expectations of what should be supplied by the forest. Forest policy is concerned with all the functions of the forest resource, the most important of which are:

- the production of wood and other forest products;
- the protection and conservation of the environment (soil, water, wildlife etc.);
- the provision of opportunities for recreation.

The production of *wood for industry* and, in some cases, for the export of forest products has been and is likely to continue to be the main policy objective for a large proportion of the managed forests of most countries and it constitutes the main source of income from forestry. But in many developing countries, and especially those with limited forest resources, the provision of *fuelwood* and of *other wood for domestic use* in rural areas is even more important. In these countries most people depend almost entirely on wood to cook their food and this situation is not expected to change much in the forseeable future, not even in some oil producing countries such as Indonesia and Nigeria. The numerous *forest products other than wood* range from edible berries and fungi to cork and from resin to fodder for cattle and venison. Where such products are of economic or social importance, even locally, they may become the concern of forest policy.

Virtually all forests everywhere have an *environmental function*, but the nature of that function varies. In major water catchment areas forests act as a sponge that absorbs rain and then releases it slowly thus helping to prevent the periodic flooding and drying out of rivers; on slopes forests help to retain the soil and prevent erosion; in arid regions forests impede the march of the desert; and almost everywhere they are playing an increasingly important role as habitats for wildlife and as *in situ* gene banks for plants and animals. In some industrialized countries the role of forests in the conservation of traditional landscapes has also become a matter for public concern. Almost invariably, the costs of conservation measures, whatever their nature and objective, are incurred on particular tracts of forest while the benefits accrue to the community at large; moreover such measures are usually only effective if taken over a sufficiently large area. Forests can therefore only fulfil their environmental functions if these are made a prime concern of forest policy. This implies that forest owners should not be expected to bear the whole of the extra costs and loss of revenue associated with meeting their environmental responsibilities. Grants and tax concessions to forest owners in this context are payments for services rendered rather than subsidies.

The use of forests for recreational purposes has only in recent times created problems requiring policy decisions. The most serious problems arise in developed countries, which are densely populated and have only a limited area of forests. Under these conditions the number of people wishing to visit forests is very large and the risk of damage great unless appropriate measures are taken.

Policy decisions refer in particular to the right of the general public to enter forest land and, where such a right exists, to the facilities, if any, which should be provided (footpaths, lavatories. picnic sites etc.). It is rightly accepted in most countries that access on foot should be permitted subject to certain conditions, but that no special facilities may be installed without the owner's permission and that he should not be forced to contribute to their costs.

The list of functions mentioned in the above is far from complete. Furthermore, attention must be drawn to possible conflicts between the different functions in the same forest. Although multipurpose management is usually possible within certian limits any one of the three mentioned major functions may conflict with another one. Preserving a forest for the purpose of gene-conservation will necessitate the reduction or complete cessation of timber extraction. Weekend houses cannot be allowed in a nature reserve, and so on.

In order to detect such conflicts in time, it is useful to translate the function or the functions of a certain forest area into goals or objectives as well as into management decisions (methods, equipment, personnel etc.) belonging to those objectives. In that way the consequences of particular courses of action are clarified. It follows from what has been said above that it is an important aspect of forest policy to ensure that the decisions taken in relation to the various functions should be consistent with one another. This elementary consideration is often ignored. A common example is that an exaggerated policy of conservation may not only unnecessarily interfere with policy measures aimed at increasing timber production but may even contribute to the gradual destruction of the forests to be preserved. Insufficient fellings may lead to an accumulation of over-mature trees which should be replaced by young ones; moreover, where the people living in and near a forest cannot derive sufficient benefit from it either in the form of gainful employment, or from the sale of produce if they own woodlands, they may even set fire to the forest in order to use the land then for grazing. Conversely an indiscriminate policy of afforestation has sometimes quite needlessly interfered with the implementation of policies for the conservation of habitats of wildlife. Many more examples could be quoted, but these suffice to illustrate the dangers of piecemeal decisions on forest policy and the need to integrate forestry into an overall land use policy.

1.3.2 Forestry, the art of managing the forests

As long as forest is abundantly available for several purposes, and the forest use, especially the extraction of products, is limited, the forest resource as an eco-system will be able to sustain itself. Above a certain level of use, however, the ecosystem cannot do so and the very existence of the forest resource is endangered. If the remaining forests are sufficient from the point of view of quantity, area, and quality, in relation to the expected forest functions, the conversion of some forest land to other use, e.g. grazing may be justified. In modern times *land use planning* will offer a tool to decide upon such an issue.

On areas to be retained as forest, the management of the resource is essential. The simple extraction of products must be replaced by a set of goal-oriented activities which ensure the future of the resource and the sustained yield of forest products or services or of both with the object of achieving optimal human benefits.

In Europe, several hundreds of years ago, such sustained yield forestry was developed. In the beginning it was especially aimed at wood production, but gradually the other forest functions were incorporated. During colonial times, this type of forestry was introduced into other continents as well. Nevertheless, in some regions of the world the 'mining' practices have continued into present times.

Of particular concern are the tropical forests, where there has been an alarming and continuing decrease of forest area in recent decades. However, in the tropics, the European and North-American experiences with sustained yield forestry have to be translated into terms of other forest types, sites and socio-cultural settings. Many of the tropical rainforests are as yet unexplored and the knowledge about these ecosystems and suitable methods of managing them are far from complete. It is important to note that there is no common denominator for the quantification of the several functions: the production functions are mostly quantifiable in terms of money but some of the service functions are not as will be explained in more detail in chapter 4. For example, a landscape or a wildlife habitat may be of immense value but they cannot be 'priced'. Methods such as multicriteria analysis and other modern approaches to decision making as employed in economics can assist in the search for practicable solutions when choice problems are involved.

As explained in the foregoing section, not all functions and, as a consequence, not all objectives in one and the same forest can be reached independently from each other; and it is a fallacy to assume, as the so-called 'wake-theory' does, that forest management focussed on timber production will necessarily ensure or facilitate the achievement of the conservation and recreational functions.

The extent to which it may be practicable, by careful planning, to combine timber production with environmental objectives and the provision of recreational opportunities under European conditions, was examined by Madas and Coulon (1975) and similar conclusions are likely to apply elsewhere. The more important points are summarized below:

In the framework of the three functions the multiple and multi-purpose use of forests is sometimes possible. In a number of cases the use of forests for a given function may not be in conflict with its use for other functions. Under other conditions, one use has a directly negative effect on other uses.

The point to make is that every forest resource has a certain capacity, at a given level of management and investment, to provide and to be the limit of a combination of functions. Once that capacity is exceeded, deterioration of the environment will ensue, and a progressive lowering of the resource's capacity.

The principle and objective of comprehensive forest policy and management should be to ascertain and attempt to provide the optimum combination of functions – optimum in relation to actual and prospective demands from society – and to restrict as much as possible the conflicts between the different functions.

Terms used to express this concept include multiple use, or balanced use or multi-purpose forestry, and integrated resource management, among others.

Among the important and difficult questions which have to be dealt with in this connection are:

– assessing the capacity of the forest resource for a given combination of functions, obtaining public acceptance of the capacity limitations, and following management practices designed to provide the optimum, but not to exceed it;

– arriving at a political consensus whether the conservation and recreation functions should be subsidized in whole or in part, and on how they should be subsidized, and by whom, and determining a reasonable basis on which to calculate the level of subsidization;

– arriving at the recognition that the forest resource is part of the national wealth; that there are the rights and the obligations of present and future generations in respect of the use of this wealth; and that ownership, including private ownership, carries with it certain obligations to society as well as the right of fair compensation.

The production function is clearly the primary use of most forests.

Single (or exclusive) use of the forest resource rarely exists, primary use is common, multiple use is probably gaining in importance, especially in forests near population centres.

Despite encroachment by the other two functions, wood production is likely to remain the primary function this century, with the exception of countries or areas with high population densities and mountainous or other difficult types of terrain, where the importance of the recreational and infrastructural services of the forest is already important and will become even more so.

The decision on the optimum combination of the particular functions of forests is a political concern and therefore one of the main topics of forest policies, but political decisions must have a sound motivation induced by experts, in this case by foresters. If foresters want to improve the existing forest policy in a given country, they have to initiate an action by putting forward a comprehensive conception of the optimum use of forest resources. They will be assisted in this task if they establish a constructive dialogue not only with those immediately concerned but also with the general public.

1.4 Policy aspects of forest management

Foresters are expected to manage the forest resource on a sustained yield base for optimal human benefits, but it is a matter of policy to decide what human benefits must be achieved in what places and by what means. Those decisions will lead to the kind of forestry that should be practised. Roughly speaking, four kinds of forestry or forestry systems can be distinguised:

First, from the viewpoint of nature and environment, forestry can be restricted to the preservation of forest reserves. The focus can be on undisturbed forest ecosystems or wildlife, or on the protection of the mountainous areas against erosion or avalanches. This is *protection forestry*. In some cases there must be no human interference at all. In other cases limited intervention may be permissible or even desirable. Some ecosystems cannot survive without it, for example where regeneration of the forest is prevented by excessive populations of wildlife whose natural enemies have become extinct. Outside those protection forests, however, forestry will be practised more intensively; nevertheless, a certain difference in management will appear in relation to the available sites and locations. One can consider a whole range or logical continuum from large forests, far from popula-tion concentrations managed mainly to produce timber for industry, via more or less multi-functional forests in rural agricultiral areas to small woods and clumps of trees typical of urban and industrial areas.

In the industrial countries forestry developed in step with society but the direct link between the forest and the demand for forest products or services became tenuous. The flow of timber is directed via the market and there are many intermediaries between the forest and the users of forest products. Based on that phenomenon, the forestry system in those regions is developed into an *industrial forestry system*. However, this kind of forestry which is linked with a whole sector of forestry activities, is not the only possible one.

In many rural areas, another type of forestry should have attention, in recent times called *social or community forestry*. In these areas, the direct links between the forests or their three-components and the people who live in and near them are still evident and should be maintained and developed. The forestry system must be adapted accordingly though always based on the sustained yield of products and services. The fact, that roughly half of the world population depends on wood for energy and other subsistence contributions from the forest underlines the need for this type of forestry and of *agroforestry*, which combines the growing of agricultural crops and trees on the same piece of land. On the other hand, the management of forests in urban and industrial areas, including trees in towns and along motorways presents quite different problems. Awareness is growing, that *urban forestry* with its emphasis on landscape, outdoor recreation, reduction of air pollution and so on deserves serious attention and should be distinguised though not wholly separated from the other forestry systems. One has to think not only of forests in the vicinity of cities as for example in Norway, Denmark, Germany, France and the Netherlands, but also of the concept of bringing nature into the cities themselves by more tree planting. Whether the forest administration is in charge of urban forestry or not, it has expertise to contribute.

Though it is partly a matter of professional skill and partly an art to develop and to apply the several forestry systems, it will be evident that it is a matter of forest policy to decide the appropriate forestry system for each set of circumstances – human and physical. The size and nature of the national forest area which is needed to strike a reasonable balance between society's demands for forest products and services on the one hand and for other uses of the land on the other hand thus depends on many factors which will vary from country to country. The *percentage of land area under forest* which is desirable will vary accordingly. Foresters in the past have sometimes asserted that every country should aim to have, a forest cover of 25-30%, but assertions of this kind serve no useful purpose. Such an average figure would obviously be unrealistically high for countries like the Netherlands or the UK and disastrously low for countries like Finland or Sweden. What is, however, certain is that *in most countries of the world some increase in forest area would be in the national interest and a decrease would not.*

The silvicultural systems to be applied should mainly be left to the decision of forest owners and managers, but there are also policy implications. *Plantation forestry* is usually based on the planting of one or a restricted number of tree species outside existing forests or following the *clearcutting* of smaller or larger areas of existing forest. Plantations lead to even aged stands which contrast with the stands of trees all ages produced by selection forestry which is based on the harvest of single trees or groups of trees followed by natural regeneration or enrichment planting. Various silvicultural systems which are intermediate between these extremes have also been developed. Even-aged plantations are generally easier to manage than selection forests and they produce timber in a shorter time and sometimes at a lower cost, but the quality of the timber may be lower, they offer less protection against water-erosion, provide fewer habitats for wildlife, are less pleasing in the landscape and tend to be more susceptible to damage by pests, wind and snow. For these rasons clearcutting is prohibited in some mountainous regions, eg in Switzerland.

The *investment implications* of forestry also must be considered at both local management and *at government policy level*. Nationally, a country's financial resources, its balance of payments, competition for resources – land, manpower as well as finance – have to be considered. In a national budget forestry may, for example, have to compete with projects for agricultural development. Where such political choices have to be made, forestry tends to be at a serious disadvantage for several reasons: insufficient weight is often attached to the *unquantifiable benefits of the service functions*, the agricultural lobby tends to be stronger than the forestry lobby (forest owners and professional foresters have much to learn) and money invested in forestry may take longer to produce a return, although this does not always apply, e.g. when forests hitherto not in use are opened up by new forest roads. For *the individual forest owner,* whether public or private, an investment in forestry can be far more attractive than is commonly assumed, but as with all investments, there are also dangers especially for those who are ignorant or careless. Governments would be well advised to make the investment implications of forestry better known.

A major task of forest policy is to settle the priorities and the *right balance between long term* and *short term objectives* and *between* measures to satisfy *local rural needs* of wood for fuel and other domestic use on the one hand and producing *wood for industry* and primarily *urban use* or *export* on the other hand. It is now widely recognized that national forest policies especially in poor countries have tended to place too much emphasis on long term industrial (urban) objectives and too little on satisfying short term urgent needs of people living in and near forests. There are welcome signs that this trend is gradually being rectified.

Forests can only survive and fulfil their role, if forest users and forest owners are prepared to take a *long term view* of forest management; but few people outside forestry do take a long term view in daily life. For that reason, forest owners and foresters occsionally get out of step with the rest of society and withdraw behind their trees; this is unfortunate: *foresters are part of society* and should act accordingly. Only in that way, can foresters contribute, to a better knowledge and understanding of forestry by the public and by those who form public opinion. Of course, foresters must be prepared not only to talk and teach but also to learn and listen.

Forests suffer from *illegal activites.* Those motivated by greed and corruption are a threat to forests more widespread and less understandable than infringements of the law to satisfy the basic human needs of the poor.

Indeed, forests are easy to plunder. The theft of Christmas trees is a common example in the Western societies. Poaching is another deep-seated evil, ranging from sports-like activities to the ruthless destruction of whole animal populations. At least as serious as poaching is the *illegal felling* of trees and even parts of whole forests for commercial purposes. In most countries such illegal fellings are punishable by law, but they continue. Illegal fellings are neither compatible with sustained yield nor with long term forest development.

In the case of a subsistence issue, e.g. poor populations living in forest surroundings, the remedy for illegal fellings are measures which enable the people concerned to obtain the necessities of life without having to break the law and without destroying the forest. This is an important reason for forestry for community development.

Corruption is a canker of public and commercial life in many parts of the world. In forestry, the felling, extraction, transport and sale of forest produce offer many opportunities for corruption at all levels of management. Corruption has been responsible for forest devastation on a vast scale and the evil is difficult to eradicate without removing the underlying causes, among which unreasonably low remuneration and the lack of a 'work ethic' or 'professional pride' deserve special mention. The elimination of complicated bureaucratic procedures involving too many forms and permits may be a useful first step in tackling the problems. Experience in many countries has shown that exaggerated control procedures, far from achieving the purpose of protecting the forest, merely make it easy for those who operate the system to exact bribes.

Beside the establishment of sensible laws and regulaations, and a competent forest authority to administer them, one of the most effective approaches to overcome the negative developments mentioned above, is *to make everybody understand that forests and trees are essential to safeguard the welfare and wellbeing of all members of a society*. Forest policy must provide the framework for this highly important field of activities.

1.5 Forestry in relation to other policies

Basically, forest policy deals with:

– the use of land for forestry including all secondary activities;
– the provision of products and services, the underlying aim being to balance the demand and supply of forest products and services.

Both these aspects affect the relations between forest policy and other policies. As already stated, the practice of forestry is not confined entirely to forest land nor to the management of forests and trees alone, although these contribute what might be called the central core of forestry.

The many different activities include, among others; site protection, promotion of the most profitable tree species (eg selection, breeding and even cloning), thinnings and the harvest of mature trees; working methods, ergonomics and organization; development and use of tools and machinery; forest inventory and planning. However, as soon as the raw material is delivered for transport on the forest road or for further sorting at a storage yard, another broad field of activity comes in view which concerns not only forest policy but also *transport policy, industrial policy* and *trade policy*. In many instances these activities are shared with the handling of imported raw materials from elsewhere and include the transportation to the mill or other plant; the follow-up of processing the raw material into products for direct use or for further handling in industrial processes such as the manufacture of sawnwood, veneer, plywood, pulp and paper, or other products like resin, oils and other chemicals derived from forest products. An endless range of products could be mentioned; in many, the original forest raw material is difficult to recognize.

The forest based activities are widened farther if we add the printing of newspapers, books, administrative forms, etc. or the use of primary forest products (sawnwood etc.) in buildings, railroads, ships, cars and even planes or for the manufacture of household furniture. All these end uses are linked to the forest and the land on which it stands by a series of activities which together constitute the forestry sector of the economy.

Forest policy thus takes its place beside other sector policies for natural resources, structured in a similar way, such as agriculture, mining and the like. The relations between such policies can easily be recognized now. Roughly spoken, they are twofold.

First, the above sector policies are all based on the use of natural resources. From the balance between the demand for forest raw material and their supply ('home grown' or imports), the need for woodlands or forests can be derived.

A certain competition between the several types of land use will arise as soon as land becomes scarce. The same holds for competition within one sector in the case that forest land use can be applied in different ways, eg. wood for pulping in competition with wood for high quality sawnwood or veneer.

Competition for land between agriculture and forestry is common. However, there is much land which is highly suited for forestry but only of marginal value for agriculture and sometimes kept for agriculture or animal husbandry by subsidies for larger than would be needed to convert such land to forestry. There are some examples of this kind of situation in the EEC. Competition arises also in the case, that the produce of one sector can be substituted for the produce of another sector, eg fossil fuels versus fuelwood. A coherent *land use policy* provides a sensible basis for defining the relations between forest policy on the one hand and other sector policies involving land on the other hand.

Forestry and agriculture, however, also have many interests in common. For example, woodlands provide farmers not only with useful produce but also with shelter for agricultural crops (wind breaks etc) and additional employment and earning opportunities at times of the year when there is little farm work. Conversely, where there are no farming communities, the recruitment of forest workers may be difficult. This point will be elaborated in the section on farm forestry in Chapter 5.

Forest policy, not only has points of contact with other land use policies. *A country's economic, social and environmental policies influence and are influenced by forest policy* in a web of complicated and sometimes not very clearly visible interactions which change as societies and governments shift their priorities. Thus in recent years, there has been an increasing emphasis on *environmental policies*. The relations between forest policy and various aspects of these other policies must be taken into account if inconsistencies and conflicts are to be avoided. The next part of this section will touch on a few of the more important of these relations.

The basic aim of forest policy that forests should be managed and used for human benefits provides the link between forestry and society. Society has various options with regard to the products and services of the forests. The forest functions and the management goals derived from them have already been discussed. Here we must note that any forest function can be linked with other policies, even with policies which are not directly based on land resources, such as those concerned with economics, social welfare and with cultural and environmental matters. *Society is therefore not only the consumer of forest products and services but is also strongly interested in the means by which its interests will be met.*

The links with *economic policy* are perhaps the most important because economic considerations largely determine the way in which choices are made by society, generally in response to scarcity of any products or services including those provided by forests. Macro-economic considerations will largely determine the quantities of the forest output in any nation and its contribution to GDP (Gross Domestic Product). However, part of the output of forestry cannot be valued in terms of money, but covers also the so-called unpriced benefits, such as soil protection, wildlife, and so on. In economics they belong to the so-called positive 'externalities' and forestry has a lot of them. Therefore, in most cases forestry is undervalued in the GDP figures; the undervaluation becomes even more serious if also some of the commodities themselves are left out of the reckoning as in the case of much fuelwood and other wood for domestic comsumption gathered by local people for their own use. Conflicts may arise between the micro-economic interests of individual forest owners and the macro-economic interests of the nation. It may pay the forest owner to use firewood in preference to other fuels and even to sell wood as firewood to domestic consumers rather than as pulpwood to industry. For the national economy, however, it might be more profitable to convert the wood into other products, e.g. pulp and paper or particle board, thereby offering more employment and more products for export. Moreover, processing the wood may pay for the import of even a larger volume of oil than is saved by using wood as fuel. Sweden is an example where such conflicts of interest have recently occurred.

Economic considerations also play a major part in determining the most appropriate balance between various forms of land use. The demand at home and abroad for timber, and for agricultural produce including non food crops such as rubber, sisal and oil palms, as well as the employment and income generated by each and the investment required, must all be considered not only in general terms but also in relation to the different types of soil and site that are available. A crucial economic factor is the very high free market price of land for urban and industrial development. Where such land is forest or in agricultural use, owners will be very tempted to sell unless government places restrictions on conversions to urban use. Such restrictions may be in the public interest for environmental reasons.

In this context *employment* must be considered. Compared with intensive farming, forestry is generally labour extensive; however, if the multiplier effect i.e. the processing, transport and marketing of the products is taken into account, the image can change considerably. Regional differences must not be neglected whenever employment aspects are considered. Forestry and the industries connected with it create gainful employment in poor rural areas where it is needed most.

This way of thinking leads to another aspect, *the social and cultural behaviour* of populations. It not only raises the question to what extent populations are aware of the benefits of forests. Other factors must not be ignored, e.g. that the consumption of forest products by urban populations far from the forests can constitute a heavy burden on the forests.

The consumption per capita differs widely between various parts of the world not only because of contrasts in the standard of living and the size of the forest resource but also because of differences in the behaviour and the habits of the people. In this way the culture of any society plays its own role and should not be neglected. This is another reason why forest policy and social aand cultural policies must be consistent with one another.

The aspect of *the environment* has already been touched upon. Of course environment raises wider issues than forestry, but forestry is closely concerned. Even where the protection of the environment is not the prime objective of forest management it places constraints on the extent to which and how the economic objectives may be achieved. In extreme cases there may have to be no harvesting of timber at all in order to achieve the environmental objectives (erosion control, nature preservation etc.) Even under these circumstances forests are of an economic value which can be estimated from the damage that would result without the forests. Some forest industries, especially pulp industries, may harm the environment by causing water and air pollution unless adequate precautions are taken. A difficulty that arises here are the international implications which are towfold. In the first place, the precautions are expensive which means that, unless they are taken by all countries, those that take the precautions will be at a disadvantage compared with foreign competitors who omit to do so. Secondly, air pollution generated in one country may do most of the damage elsewhere.

1.6 Co-ordination of policies

Forest policy in general is a sector policy, linked with other similar sectors, but forest policy is also subject to aspect policies which are principally cross-sectoral. The potential of a single forest, to offer a series of benefits to society, predestines forestry to be influenced by sectoral and cross-sectoral policies. Both the sector and the aspects are linked with the nature of forestry. In practice, the sectoral policies are much more developed that the aspect or cross-sectoral ones. In a sense, there are two policy 'worlds': on the one hand the 'world' of the differentiated sectors such as forestry, each well-developed with a high degree of independence; on the other hand the 'world' of the aspect network aiming for co-ordination and even, where possible, for integration, which is rather complicated.

In all these relations with other policies, those responsible for forest policy should seek to ensure that adequate weight is attached to the welfare aspects of forestry and that these are not subordinated to narrow purely financial and economic considerations. This is sometimes not an easy task because so many politicians and senior administrators understand so little about forestry and care even less.

Regional planning, which has been practiced in several countries for some decades provides a framework for the integration of sectors and aspects; it may operate at several levels. The emphasis lies on the co-ordination of all major activities in the region including those concerning land use, economics, employment, environment and social/cultural aspects; The local setting of the various items and the involvement of the local population are prime considerations. All actions together are designed to achieve the optimum use of the available resources in a part of the country: e.g. a province, a community, a watershed. In one way or another, the ultimate goal of regional planning is to balance the activities of society, based on and (vice versa) influencing the several sectoral and aspect policies concerned.

Regional planning is the key to a sensible co-ordination of policies, because every region has some characteristics and problems in common. Co-ordination is much more difficult without such common factors.

References

Coulon, M. de (1977) Politique forestiers et gestion des Forets. In: Journal Forestier Suisse.

Duerr, W.A. et al. (1979) Forest Resource Management; decision-making principles and cases W.B. Saunders Company, Philadelphia/London/Toronto.

James, N.D.G. (1981) A history of English Forestry. Basil Blackwell, Oxford.

Mantel, K (1980) Forstgeschichte des 16. Jahrhunderts. Paul Parey, Hamburg & Berlin

Sharpe, G.W. et al (1976) Introduction to Forestry. McGraw-Hill, New York.

Sinden, J.A. et al (1978) Unpriced values Decisions without market prices. John Wiley & Sons, New York.

Thirgood, J.V. (1981) Man and the Meditteranean Forest. Academic Press, London.

Worrell, A.C. (1970) Principles of Forest Policy. McGraw-Hill, New York.

2 The world perspective

Tim Peck

2.1 Introduction

'The forest is a resource already under pressure from billions of people, striving to lift their standards of living. Since the last Congress, hundreds of millions more people are having to depend on an ever-decreasing area of forest. This situation places on governments and on their forest administrations an even greater responsibility than ever before to arrest degradation and to manage every hectare of forest to best advantage and in the best interest of all people. The Congress *recognized* that this was a major challenge to the forestry profession, and to all those in any way concerned with forestry activities.'

This statement forms part of a remarkable document known as the 'Jakarta Declaration', which summarizes the conclusions and recommendations of the Eighth World Forestry Congress, held in Jakarta, Indonesia in October, 1978. It also provides a convenient starting point for this chapter, in which we shall discuss the relevance of the world forestry situation and the trends in forestry and forest products markets to the formulation and implementation of forest policy.

For the most part, major policy decisions relating to the forest aand forest industry sector are taken at the national level, but some are taken regionally, locally or by the enterprise. Forest policy formulation is practically non-existent at the international level, although it is frequently discussed at international meetings. Policy within a country, however, can seldom be made without reference to international developments. To take an obvious example, a decision to extend the area of productive forest in a country, or to establish a major new forest industry, must take into account, amongst many other factors, the prospects on the international forest products markets.

Very few countries can afford the luxury of having a forest and forest industry sector that can be kept insulated from international influences. Whether as importers or exporters – and the majority of countries are both – domestic markets are affected, through the price and other mechanisms, by international factors. These factors have an impact both on the short-term market and, what is of more concern from the policy-making point of view, on the long-term development of the national forestry and forest industry sector.

International trade in forest products has expanded in importance relative to production and consumption; and the number of countries involved in international trading, whether as exporters or importers, has also grown. To take a concrete example in the Pacific-Basin, it is certainly of signifance to an importing country such as Japan or exporting countries as Chile and New Zealand to be aware of the forest resource situation on the West Coast of North America and the policy decisions which Canada and the USA have been taking to move from 'timber-mining' towards sustained yield management; equally to be aware of fundamental shifts in trade policies of countries such as Indonesia or the Philippines from being log export oriented to a policy of developing domestic sawmilling and plywood capacity, partly with a view to replacing exports of logs by the higher unit value products of those industries.

Nor is the international influence limited to purely commercial considerations. A striking feature of the past decade or two has been the growth in concern on an international scale for environmental protection, as exemplified by the attention being drawn to the dwindling of the tropical forest resource, and its impact on local and global ecosystems.

The purpose of this chapter is firstly, to discuss the type of information on the international situation and trends and prospects which is required for policy-making in the forest and forest industry sector. Secondly, a brief global overview will be given which is intended to provide a background to the discussions in later chapters. Thirdly, some general conclusions will be drawn regarding the way in which international developments need to be taken into account in national policy formulation.

2.2 The forest resource

Many books and papers have been written describing the forest – of a country, a region, a continent. Each account is flavoured or influenced by the purpose for which the description is being made, whether botanical, economic or other.

A sound forest policy cannot exist without an adequate information base, especially on the forest resource itself. In shaping policy, it is also necessary to understand the forest resource from many points of view, but unfortunately data collection is expensive, and it has often not been possible to justify such expenditure in terms of the economic benefits obtained. Hence, for many parts of the world, the quality of forest inventory data still leaves much to be desired.

It is clearly impossible to give a detailed description of the forest resource on a global scale in a book of this size. It may nevertheless be helpful to list what appear to be the main descriptive features and to give some general information, in so far as it is available.

The last qualification deserves an explanation. Forests cover nearly one-third of the land surface of the world. However, even with the increasing use of sophisticated technology, including remote sensing, satellite photography and so on, knowledge today about the forest resource, especially in the more remote areas, is still far from complete. There are a number of reasons for this:

1. the need for detailed surveying is not great for those forests which are likely to remain out of economic reach for the foreseeable future. This applies to quite a large area in total, notably in parts of Siberia, the Amazon and Congo basins, the Canadian far north, etc.;

2. the forest, especially the tropical rain forest, is the most complex terrestrial ecosystem and is extremely variable in composition. This creates problems in the identification and listing of the vast numbers of living species of fauna and flora and the other components that make up a 'forest' (trees may be the dominating feature, but numerically they are relatively insignificant);

3. in order to communicate about, make use of and manage the forest resource, man needs to be able to incorporate it into a descriptive classification system. Despite the efforts of international organizations and experts from many countries, universally acceptable definitions, on which to base a common system, are still lacking. For example, one of the most basic distinctions should be between forest land and other land, something which on the face of it might seem straightforward. In fact, however, this has proved to be surprisingly difficult to agree on – where, along the gradient from dense forest to pure grassland, to draw the line. The foresters may, and often do, disagree on this point amongst themselves, as well as with those in other disciplines, such as surveyors or agriculturalists, as official statistics from different sources within the same country often show.

Certain broad descriptive features of the forest resource are found in most countries' forest inventories and hence in the collection of forest inventory data at the international level, by FAO in collaboration with other agencies. These features, all of which are important from the policy formulation point of view, include:

– area of forest and other wooded land in total, separated from other main land use categories;
– distinction between dense or closed forest and other wooded land;
– the main physical types of forest – high forest, coppice, bamboo forests, mangrove forests (in some cases, this distinction is dependent on the type of management or silviculture practiced, see below);
– the predominant species groups – coniferous, non-coniferous, mixed. This may be further elaborated according to the predominant individual species or groups of species;
– age-class or size-class distribution of even-aged stands and separation of even-aged and uneven-aged stands;
– distinction between 'natural' and 'artificial' stands. Sometimes such a distinc-

tion is clearcut; in other cases far from being so. In Europe, for example, it can be argued that natural forest, i.e. virgin forest undisturbed by man, has almost ceased to exist. However, man has the capability of creating and maintaining, by suitable silvicultural methods, a forest which appears quite natural;

- within the 'artificial' forest group, distinction between afforested and re-forested land, the former being stands established on land not formerly carrying a forest crop (at least not in the immediate past);

- distinction between types of ownership, a particularly important aspect from the policy point of view. The two broad categories are publicly- and privately-owned forest land, but further sub-divisions, e.g. publicly-owned into State-owned and other categories, are also important;

- the extent to which the stand in a 'dense' or 'closed' forest is stocked (stocking density), which provides an indication of the degree to which the productive capacity of the soil is being utilized by the tree crop;

- the quality of the present or future crop based on site classes, yield classes or other measures of site productivity. It is usually important also to have information on soil properties;

- the extent to which the forest resource is put to use. The concepts and terminology are by no means standardized here: forest in use/not in use; productive/unproductive; exploitable/unexploitable; operable/non-operable; accessible/inaccessible. One of the major difficulties has been of setting a distinction at a particular instant of time. What is not in use/unexploitable/inaccessible today may be brought into use or become exploitable/accessible tomorrow or in the foreseeable future as a result of developments in technology or an increase in wood prices relative to harvesting costs. Conversely, an area that is today being exploited may tomorrow become unexploitable as a result of a political decision to create a nature reserve or wilderness area. Note that it may have become 'unexploitable' for logging, but it could still be economically and physically 'accessible' and certainly 'in use', although by a different clientèle. Hence the terminological dilemma.

The type of inventory data outlined above will go a long way towards satisfying information needs at the policy level. These needs will nevertheless differ from those at the local management level.

The above descriptions apply basically to the land and vegetative cover. Some of them apply equally to the trees themselves, about which information is needed concerning the standing volume (growing stock) and increment for the total stand (or forest or country) and per unit area. Thus, inventories provide information about growing stock and increment, according to type of forest, species, age or dimension class and sometimes ownership. The concept of standing volume or growing stock is not standardized, however. Usually, it relates to what is considered to be the part of tree (trunk and main branches) which can be used as industrial raw material or fuelwood. There is lack of conformity, however, as to what constitutes the useable portion, e.g. the minimum top diameter. The increment is also based on the same concept of useability.

The coverage of growing stock and increment data also tends to vary from country to country, ranging from all trees growing on forest and other wooded land as well as trees outside the forest to only those trees in exploitable closed forest or productive forest.

As a counterpart to increment, information is needed on the drain on the forest. Drain may be defined as the withdrawal of wood from the standing volume, whether by fellings or mortality. Mortality may be considered as the difference between gross and net increment; and the difference between net increment and fellings shows whether the standing volume is increasing or decreasing. In a mature, unexploited forest, mortality about equals gross increment, so that net increment is nil. In a young, properly thinned plantation, mortality is negligible, so that net increment almost equals gross increment; and until the stand becomes mature, its standing volume will increase, except for temporary checks when it is thinned.

Until fairly recently, foresters and forest industries depending on the forest for raw material were almost the only people needing detailed forest resource information. Hence, the emphasis in inventories was on the commercially interesting part of the tree, and the exploitable areas of the forest. More recently, three developments have resulted in a significant broadening of the groups of people interested in the forest:

1. with increasing population and consequent pressure on the land, regional planners and others concerned with the allocation of land to different uses have pointed to the necessity of integrating policies for and management of forest land into overall land use policies and management; specific information on the forest land is then needed.

2. partly linked with 1. above, the environmental movement has helped to expand public awareness of the non-wood production functions of the forest. Efforts to evaluate the relative importance of the various functions – protection, wildlife, hunting, recreation and so on – have exposed the difficulties, partly through lack of inventory data on such aspects, of determining the optimum combination for the multiple use of the forest;

3. the energy shocks of the 1970s have renewed interest in biomass as a source of fuel, leading to the need for information, previously lacking, on the total volume of forest and tree biomass from which to access its potential contribution to future energy supplies.

Considerable research work is in hand in several countries to extend the scope of forest inventories to incorporate the new types of information needed for the above purposes. FAO, ECE and other agencies are playing an active part in stimulating interest in these questions and have started to collect such information at the international level. It will be some years, however, before their efforts can be expected to bear fruit.

In the meantime, we must continue to depend on the more traditional type of inventory information available. Even so, as mentioned earlier, there are wide differences between countries in coverage, detail and reliability. Whatever the shortcomings, inventory data on a worldwide and regional scale have to form part of the information base needed for national forestry policy formulation.

2.2.1 Land use categories

Of the total world land area of approximately 130 million square kilometres (km²) forest and woodland is estimated to account for just under one-third (31%), or 41 million km². This compares with over 45 million km² of agricultural land (35%) and 44 million km² (34%) of all other categories, the bulk of which is 'waste' land (desert, tundra, etc.), but which includes built-on land. The area of built-on land is not known, but it is increasing at the expense of other categories.

As table 2.1 shows, the proportion of forest and wooded land is around the one-third mark in most regions[1] of the world. Three regions – the USSR, Latin America and Japan (perhaps surprisingly) – have a forest cover percentage well above the world average; and three well below – the Near East, China and the group 'other developed countries'.[2]

The total area of forest and woodland is only a partial indicator of the importance of the forestry sector to a region or country. Converting it into terms of area per inhabitant helps to fill this gap. At the present time, there is on average in the world slightly less than one hectare of forest and woodland per caput. This area has been decreasing and will continue to decrease under the twin influences of rising population and, in some regions, declining forest area. It has been estimated that 50 years ago, in the 1930s, there were two hectares per inhabitant; today there is less than one, and by the year 2000 it will be only half a hectare per person.

It is in the developing regions that the area per caput is declining most noticeably; for the developing world as a whole, the area per person of 0.7 ha/caput is already less than half that in the developed world (1.6 ha/caput), and in China it is as little as 0.1 ha/caput – a factor which has surely had a strong influence on the emphasis being given to afforestation.

The regions best endowed with forest and woodland, in terms of area per caput, are the USSR, Latin America, North America, Africa and 'other developed and developing countries'; and the least endowed, apart from China, are Japan (despite its high percentage of forest cover), Europe and 'other Asia/Pacific'. The last group of countries, which excludes China and Japan, has 9% of the world area of forest and woodland but as much as 30% of the population. Adding in China and Japan, the shares are 12% and 54% respectively.

The figure of area per inhabitant still does not tell us enough about the capacity of a country's or region's forest to meet the needs of the population for wood and non-wood forest products. This depends, amongst other things, on the quality of the forest (and the way it is managed) – species composition and type of tree – the volume of wood, its location in relation to the main consumption centres and processing industries and the increment which can vary enormously according to species and site.

2.2.2 Main types of forest[3]

It is not easy to define a coniferous, non-coniferous or mixed forest on the basis of area: the dividing line between them is bound to be arbitrary. However, the world's forests can be identified as falling into these three groups. The coniferous group is predominant to the north of the Tropic of Cancer and at higher elevations, but there are also extensive areas of pines and a number of other coniferous species in tropical and sub-tropical areas.

Roughly three-quarters of all coniferous[4] forests lie in the coniferous belt which separates the northern tundra from the temperate-zone forests. The belt stretches across Alaska and Canada, northern Europe and the northern two-thirds of the USSR. While the forests covering this enormous area are similar in structure, the tree species differ. In northern Europe and north-western USSR, Norway spruce (*Picea excelsa*) and Scots pine (*Pinus sylvestris*) predominate. Further east, this association is replaced by Siberian fir (*Abies sibirica*), larches (*Larix* spp.) and Siberian pine (*P.sibirica*). In the North American section, the species to the west are lodgepole pine (*P. contorta*), sub-alpine fir (*A. lasiocarpa*) and black spruce (*P. mariana*), giving way to the east to larches, white spruce (*P. glauca*), balsam fir (*A. balsamea*) and jack pine (*P. banksiana*).

The coniferous montane forests are a feature of the great mountain ranges – the Rockies, Cascades and Sierra Nevada in North America, the Alps and Carpathians in Europe, the Himalayas and the Hindu Kush, and in the southern hemisphere, the Andes. The number of tree species in these forests may be considerable, due to the vertical zoning as well as the marked differences in rainfall between one side of the range and the other. The diversity is greatest in North America, partly because the ranges run north-south so that latitude has an influence as well as altitude. To name a few of the major tree species in these North American forests: Douglas fir, lodgepole pine, western white pine, Engelmann spruce, ponderosa pine and western larch. This is also the region of the giant coastal redwoods and sequoias (*Sequoia sempervirens* and *Sequoiadendron giganteum*), although the former is restricted in its natural state to the lower-lying coastal areas of California. These montane forests combine to an almost unique degree the multiple functions of the forest – wood production, protection, recreation and other social services.

Two other types of coniferous forests should be mentioned: the lowland pine forests which are found, according to the genus, in both temperate and tropical

Table 2.1 Land use in 1980

	Land area (excl. water)				Land area per caput[a]			
	Total	Agri-culture	Forest & woodland	Other	Total	Agri-culture	Forest & woodland	Other
	(million ha.)				(ha/cap)			
World	13075.2	4568.9	4093.5	4412.8	2.9	1.0	0.9	1.0
Developed	5484.7	1941.3	1829.2	1714.2	4.7	1.7	1.6	1.5
North America	1834.8	496.4	610.6	727.8	7.3	2.0	2.4	2.9
Western Europe	373.1	166.5	125.5	81.1	1.0	0.4	0.3	0.2
Eastern Europe	99.7	61.2	29.2	9.3	0.9	0.5	0.3	0.1
USSR	2227.2	605.7	920.0	701.5	8.4	2.3	3.5	2.6
Japan	37.1	5.5	25.0	6.6	0.3	0.05	0.2	0.05
Other developed[b]	912.8	606.0	118.9	187.9	18.0	12.0	2.3	3.7
Developing	7590.5	2627.6	2264.3	2698.6	2.3	0.8	0.7	0.8
Africa	2331.2	784.9	641.9	904.4	6.2	2.1	1.7	2.4
Near East	1192.3	375.0	97.6	719.7	5.6	1.8	0.5	3.4
China	930.5	319.2	116.4	494.9	0.9	0.3	0.1	0.5
Other Asia/Pacific	1028.1	444.5	355.9	227.7	0.8	0.3	0.3	0.2
Latin America	2020.0	702.5	1015.2	302.3	5.5	1.9	2.8	0.8
Other developing	88.4	1.5	37.3	49.6	17.0	0.3	7.2	9.5

	Percent of total land area (%)				Percent of world total (%)		
World	100 0	34.9	31.3	33 8	100.0	100.0	100.0
Developed	100 0	35.4	33.4	31.2	41.9	42.5	38.8
North America	100 0	27 0	33.3	39.7	14.0	14.9	16 5
Western Europe	100 0	44.6	33.7	21 7	2.8	3 1	1 8
Eastern Europe	100.0	61.4	29 3	9.3	0.8	0.7	0.2
USSR	100.0	27.2	41.3	31.5	17 0	22.5	15.9
Japan	100.0	14.8	67.4	17.8	0.3	0.6	0 1
Other[b]	100 0	66.4	13.0	20.6	7.0	2.9	4 3
Developing	100.0	34.6	29.8	35.6	58.1	55.3	61.2
Africa	100.0	33.7	27.5	38.8	17.8	15 7	20.5
Near East	100 0	31.4	8.2	60 4	9 1	2.4	16.3
China	100 0	34.3	12.5	53.2	7 1	2.8	11 2
Other Asia/Pacific	100.0	43 2	34.6	22.2	7 9	8.7	5.2
Latin America	100.0	34.8	50.2	15.0	15.5	15.4	6.9
Other[c]	100.0	1.7	42.2	56.1	0.7	–	1.1

Source: FAO Production Yearbook 1981 (Vol. 35).

a Detail may not add to total, due to rounding.

b Australia, Israel, New Zealand, South Africa.

c Mostly islands in the Pacific region, but also includes Greenland which accounts for nearly two-fifths of the total land area.

zones; and the southern hemisphere conifers, including species of the genera *Agathis, Araucaria, Podocarpus*, which are scattered, often as forest remnants, over a range of latitudes from Japan and China to Chile and New Zealand. In addition to the natural stands of pine, for example in the south-eastern United States, plantations of these species, of indigenous or exotic origin, are playing an increasingly important role in the international supply of long fibre. Radiata pine in New Zealand and Chile and maritime pine in Portugal, Spain and France are just two examples.

Turning to the non-coniferous[5] forests, roughly three-quarters of them, in terms of standing volume at least, occur in a belt – less continuous than the northern coniferous belt – on either side of the equator. The farthest north that these tropical-zone forests are found is southern China and Mexico; the farthest south, northern Argentina, Madagascar and Australia. It is impossible to list the tree species, which run into thousands. These forests follow a gradient from dense tropical rain forests, where annual rainfall exceeds 2000 mm a year, to dry wooded savannah and from the cloud forests at higher elevations to the mangrove swamps on the coast.

The tropical rain forest is virtually non-seasonal, which means that as an entity it is evergreen. Some species may shed their leaves at certain times, while flowering and fruiting follow a certain seasonal rhythm generally related to the rainy periods. It has as many as five storeys, dominated by scattered emergent trees which, if of the right species, are the main target of commercial exploitation.

The heterogeneity of species and sizes and the complexity of the tropical rain forest ecosystem is both a strength and a weakness. A strength, because when left in its natural state, it is a gene bank of immense reserves and diversity. A weakness, because once disturbed by commercial exploitation, shifting cultivation or other forms of interference, some of the links of the chain of interdependency in the ecosystem are broken, and the process of degradation that ensues may become irreversible. A further weakness, which does not become apparent until the tropical forest ecosystem is disturbed, is the low nutrient holding capacity of the soil. Under normal conditions, nutrients in the litter layer are quickly broken down and recycled, and there is little chance for the minerals to be leached out of the soil by high rainfall. Forest land cleared for farm crops loses its fertility within two to three years, however, and the stage is then set for the deterioration process leading to erosion and eventual desertification. It is significant in this regard that recent scientific investigations have shown that only 2-3% of the soil in the huge Amazon basin could successfully support commercial agriculture and forestry.

Despite its lushness, therefore, the tropical rain forest is a fragile ecosystem. Because of its complexity, it is still not well understood by man, whose attempts at silviculture have often, for that reason, failed in the past.

Other special types of forest in the hotter climates are the moist deciduous

Table 2 2 Area of tropical forest and other land with woody vegetation, 1980 (million ha)

Type of tropical forest	Africa, South of the Sahara^a			Asia/Pacific			Latin America			Total		
	Total	Non-Coniferous	Coniferous	Total	Non-Coniferous	Coniferous	Total	Non-Coniferous	Coniferous	Total	Non-Coniferous	Coniferous
Natural closed forests	215.5	214.4	1 1	300.4	292.0	8 4	678.7	653.9	24 7	1194 5	1160.3	34 3
Productive	162 3	161 7	0 6	197 5	191 9	5 6	521.7	506.4	15.2	881.4	860 1	21.4
Unproductive	53.2	52 7	0 5	102.9	100.1	2 8	157.0	147.4	9 6	313 1	300.2	12.9
Natural open forests	486.4	486 4	-*	30 9	30.9	-*	217.0	217.0	-*	734.4	734 4	-*
Productive	169 2	169.2	-	8.5	8.5	-	142 9			320.6		
Unproductive	317.2	317 2	-	22 4	22.4	-	74.1			413.7		
Industrial plantations	1.0	0 5	0 5	3.5	2.9	0 6	2.6	1.0	1 6	7.1	4 4	2 7
Bamboo forests	1.1			6.2			..			7.3		
Productive	0 7			3 5			..			4.2		
Unproductive	0.4			2 6			..			3 0		
Fallow forest	166.0	166.0	-	72.2	71.2	1 0	170 3	160.9	9 3	408.5	398 1	10.3
Closed	61.6	61 6	-	68 2	67 2	1.0	108.6	99.3	9.3	238.5	228 2	10 3
Open	104.3	104.3	-	4.0	4.0	-	61 7			170.0	170 0	-
Scrub-form woodland	442 7	..	..	35.5	..	..	145.9			624.1		
Total	1312.8			448.7			1214.4			2975 9		

Source. Tropical Forest Resources, FAO Forestry Paper 30.

^a Excluding South Africa

Note. detail may not add to totals, due to rounding

The meaning of the symbols used in this table and elsewhere in chapter 2 is as follows:

- nil or negligible

* unofficial or estimated figure

.. unknown or not available

forests, where the rainfall is lower and more seasonal than in the rain forest areas; and montane forests, including at certain elevations and topographical conditions the cloud forest, where the hot humidity of the lowlands gives way to the cooler dampness of the mountain slopes, producing a markedly different type of vegetation. There are also mangrove swamps along tens of thousands of kilometres of coast and forests of bamboo which, although of the grass family, may grow to 30 metres in height. According to FAO, there are over seven million hectares of such forests in Asia and Africa.

Providing the bridge between the tropical forests and the northern coniferous belt are the temperate forests. Some of them, as mentioned earlier, are predominantly coniferous, usually pine, but the largest proportion are mixed or predominantly non-coniferous. These are, over most of the temperate area, deciduous species such as oak, beech, maple, poplar, ash, hickory etc., but in drier, hotter areas, such as the Mediterranean basin, there are evergreens – the holm and cork oaks and the olive. The main areas of temperate forests are central and southern Europe, eastern Asia and eastern North America. There are also extensive areas of temperate forests in western South America, North America, South Africa, Australia – dominated by the eucalyptus family – and New Zealand. A contrast between the northern and southern hemisphere temperate forests is that the latter are predominantly evergreen, despite similarities in the climatic conditions under which the deciduous forests grow in the north.

The above highly compressed and over-simplified description of the world's forests is based to some extent on the situation where man has not made a major impact. Where he has, the forest has often reverted to an earlier stage of succession, or through silviculture, man has deliberately kept it there. Examples are the natural regeneration by alder, aspen and birch of felled areas in the northern coniferous belt; the maintenance of a high proportion of spruce in what would probably otherwise be predominantly beech and fir forests in central Europe; the replacement of natural tropical forest by plantations of pines, eucalyptus and other fast-growing species: and of the natural temperate forests in New Zealand, Australia and Chile by *Pinus radiata*.

Man's interference with the natural forest, whether deliberately with the object of 'improving on nature' for economic gains or for environmental reasons (soil protection, etc.) or thoughtlessly – removal of the forest cover in areas where it could be restored only with great difficulty – is affecting a steadily increasing part of the world's forests. There is virtually no virgin forest left in Europe; it is becoming scarce in the United States and western USSR; many parts of the drier tropical forest are under pressure because of the constant need for fuelwood. It is becoming critical for mankind to decide – and the decision has major international implications – on a policy towards the remaining natural forests: whether to preserve them in their virgin state; if so, how much of them, where and how. The 'how' may depend heavily on policies towards intensifying management and increasing wood yields on the more accessible forests and towards afforestation, which would help to hold back commercial and other pressures on the natural forests.

The main categories of land on which forests and other woody vegetation are growing in tropical countries are shown in table 2.2. The total for the 76 countries concerned approaches 3,000 million ha, but of this less than one-third – 888 million ha – consists of productive natural closed forests and industrial plantations. Much of the remaining land under some form of forest or tree cover is, however, important as a source of fuelwood for local populations, which for the most part are dependent on wood to provide most or all of their energy needs for cooking and heating.

Table 2.3 Estimates of area of closed forest and other wooded land

	Total forest & other wooded land	Closed forest Total	of which: Exploi- table[b]	Other wooded land	Percent of total Closed forest Total	of which: Exploi- table[b]	Other wooded land
	(million hectares)				(%)		
World	3930	2860	1905	1070	73	48	27
Developed	1886	1508	990	378	80	52	20
North America	630	510	410	120	81	65	19
Western Europe	126	108	100	18	86	79	14
Eastern Europe	30	30	27	–	100	90	–
USSR	920	785	390	135	85	42	15
Japan	25	25	24	–	100	96	–
Other developed	155	50	..	105	32	..	68
Developing	2044	1352	915	692	66	45	34
Africa	563	203	150	360	36	27	64
Asia/Pacific	606	454	335	152	75	55	25
Near East	81	14	..	67	17	..	83
China	115	90	..	25	78	..	22
Other	410	330	..	80	80	..	20
Latin America	875	695	430	180	79	49	21

[a] Data in this table are taken from different sources than those in tables 2.1 and 2 2, including FAO's *Agriculture towards 2000* and FAO/ECE's *European Timber Trends and Prospects, 1950 to 2000* (ETTS III) The difference at world level is fairly insignificant, 163 ml. ha (4%), but there are relatively larger differences at the regional level, especially in 'other developed', Near East and Latin America. The problems of definition which cause these differences are discussed in the text.

[b] The concept of 'exploitable' or 'operable' forest is also discussed in the text. It should not be inferred that all exploitable forest is currently in use. There are large areas of such forest which may in fact never come into use or under management, even though theoretically exploitable.

The estimates of area of closed forest and other wooded land in table 2.3 are not comparable, for reasons discussed earlier, with the figures in tables 2.1 and 2.2. They serve as an indication, however, of the distribution worldwide of the main categories of tree-bearing land. The area of exploitable closed forest is more or less equally divided between developed and developing regions. On the other hand, the latter have nearly twice as much 'other wooded land' as the former, reflecting the importance of natural open forests and scrub woodland already noted in table 2.2 for the tropical countries.

2.2.3 Standing volume

While estimates of the standing volume of the forests of most countries in the developed regions can be considered adequate within a reasonable margin of error (perhaps 10-20%), unfortunately the same cannot be said for many developing countries. The figures in table 2.4, therefore, should be taken as being only indicative. They show that coniferous species account for about two-fifths of the world's total standing volume in closed forests and non-coniferous for three-fifths. However, the coniferous species are very heavily concentrated in the developed regions, which have over 90% of the coniferous total. Developing

Table 2.4 Approximate data of standing volume in closed forests in the 1970s

	Total	Coniferous	Non-coniferous
	(1000 million m³ overbark)		
World	301	112	189
Developed	137	105	32
North America	44	32	12
Europe	14	9	5
USSR	75	62	13
Other developed	4	2	2
Developing	164*	7*	157*
Africa	40*	–*	40*
of which: Tropical[a]	39	–	39
China	6*	2*	4*
Other Asia/Pacific	36*	3*	33*
of which: Tropical[a]	32	1	31
Latin America	82*	2*	80*
of which. Tropical[a]	78	1	77

Sources: ETTS III; Persson; FAO; Peck

[a] Countries covered by 'Tropical Forest Resources', FAO Forestry Paper No. 30.

* Estimates based on out-of-date and incomplete information.

countries, on the other hand, have over 80% of the non-coniferous standing volume.

It can be estimated that three-quarters of the world's total standing volume is concentrated in two broad forest belts. Three-quarters by volume of all coniferous species occur in the northern coniferous belt. Three-quarters by volume of all non-coniferous species are found in the tropical moist forest belt.

Compared with the highly speculative figure of about 300 thousand million m^3 of standing volume in closed forests given in table 2.4, Persson (1974) estimated that the absolute maximum volume might be 450 thousand million m^3, but considered that 350 thousand million m^3 would be more realistic. This figure included the standing volume in both closed and open forest as well as on trees outside the forest.

With regard to increment, no useful estimate can be made for natural forests, where mortality (natural losses) more or less equals forest growth. For regions where most of the forest is managed or at least in use, such as Europe and Japan, gross and net increment figures can be obtained with some degree of reliability; while for North America and the USSR, with partly natural and partly managed forests, some estimates have been made. Net annual increment for the developed regions is estimated at 2,400 million m^3, 65% of which is coniferous. It should be noted that the increment figures do not necessarily have any relationship with potential cut. Especially where mature and overmature forests predominate, potential cut could be considerably higher than net increment, if the object is to create a forest with a 'normal' distribution of age-classes.

2.2.4 The outlook for the forest resource

Uncertainties were expressed earlier about the rate at which the tropical forest is disappearing. FAO's estimates of the rate at which deforestation is occurring at the present time in the 76 countries with tropical forests is illustrated in simplified form in table 2.5. Roughly 11.3 million ha of natural forest are cleared each year, of which two-thirds or 7.5 million ha are 'closed tree formations' (closed forest) and one-third or 3.8 million ha 'open tree formations' (other wooded land). Of the 11.3 ha, rather more than half – 6.2 million ha – ceases to be forest, but 5.1 million ha do remain forested, but frequently in a degraded form. It is by no means easy to assess, therefore, precisely the rate at which natural tropical forest is disappearing.

Contrary to assumptions made by some other experts, FAO believes that the rate of loss of tropical forest to other land uses, mainly agriculture, may gradually slow down, not so much because of any easing of pressure on land, but because of the increasing inaccessibility of the remaining natural forests. On the other hand, road- and railway-building, as in Brazil and Columbia, are still opening up hitherto inaccessible tropical forests.

From the forester's point of view, changes in land use can be considered as positive, neutral or negative in impact. Amongst the positive changes are:
- conversion of low grade to productive forest. This is an important feature of European forestry. ETTS III forecast a net increase of 14 million hectares of exploitable forest between 1973 and 2000, against a net increase of 8 million ha in the total area of forest and other wooded land. Much of the difference would be the result of conversion of other wooded land into exploitable forest;
- afforestation. This is taking place in most countries of the world, usually with one of two main objectives: wood production and environmental protection. Examples of the latter are the planting against desert encroachment around the Sahara; the planting of shelterbelts to protect agricultural crops on the steppes of the USSR; the planting of green belts around Khartoum (Sudan) for the protection of the inhabitants and their crops against sand storms; planting along motorways or around airports to reduce noise. Planting for wood production is being undertaken in both importing and exporting countries, the extent depending mainly on their assessment of market needs – import substitution, export potential – and policy decisions regarding allocation of land, capital and labour. ETTS III estimated that 5 million ha would be afforested in Europe beteen 1973 and 2000, most of which for wood production;
- natural extension: the colonization or recolonization of non-forest land. This has been a recognizable development in, for example, the eastern United States, where abandoned farms have reverted to forest. A similar situation has been occurring in parts of Europe; Finland has estimated, for example, that $2^1/_2$ million ha will revert to wooded land between 1973 and 2000 as a result of natural extension.

An example of a change in land use whose impact is neutral from the forester's point of view is the transfer of productive forest (for wood) to another use, e.g. wilderness, recreation, nature reserves, etc. Some foresters might, of course, question whether the impact was neutral. Indeed, the establishment of new wilderness areas in the United States in recent times has been the topic of heated debate, with the Forest Service tending to be caught in the cross-fire between the commercial and environmental lobbies.

Changes considered by foresters to have a negative impact include:

- permanent removal of the forest cover to make way for another type of land use – agriculture, urbanization, road-building, quarrying, reservoirs, etc. It is clear, however, that in some instances policy to conserve the forest has to be weighed against other national priorities, such as the need for land for food production or industrialization. In developing countries with still extensive forest areas, the latter may very often be considered the more essential. Too often, however, deforestation may be allowed to proceed without adequate recognition of the social costs or consequences;

– degradation of closed forest. This may occur as a result of over-exploitation and lack of management to ensure proper restoration. In the northern hemisphere it has occurred where no regulations exist for the natural or artificial restocking of clear-felled areas. Considerable areas, in Canada, the USA and the USSR for example, are now covered with low-quality hardwood forests, which have established themselves after the felling of the coniferous stands. In the tropics, this type of deterioration frequently follows logging operations and shifting cultivation. Given time, such areas may become fully restored, but increasing population pressure frequently means that not enough time is given for this to happen before the site is disturbed once again.

Apart from the forecasts of a declining area of tropical forest, mentioned above, it is expected that the forest area will also continue to decline slowly in North America, at least the area available for commercial exploitation. This will be partly to provide more land for urban and industrial use, but also for wilderness areas. In Europe, on the other hand, the forest area may continue to expand, but because of increasing competition from other land uses, the process will be slow and the economic and social case for extending the forest area will have to be convincingly established. Moreover the locations in which forest areas could be expanded are often far from the main consumption centres, i.e. abandoned agricultural land in the more remote districts.

Table 2.5 Summary of shifts in land use as a result of clearance of natural tree formations in 76 tropical countries (million ha)

| | Natural tree formations | | |
| | Total | of which: | |
		Closed	Open
Area in 1980	1935	1201	734
Area cleared annually	11 3	7.5	3 8
Percent of 1980 area	0 6%	0.6%	0 5%
of which:			
Retains some form of tree cover (complex of shifting cultivation and secondary formations) ('forest fallow')	5 1	3 4	1.7
Transferred to another land use (permanent agriculture, shrub formations, brush, grasslands, deserts, reservoirs, human settlements, etc.[a])	6 2	4.1	2.1

Source: Tropical Forest Resources, FAO Forestry Paper Number 30 (1982)

[a] Conversion into forest plantations is also included here.

2.2.5 Man-made forests

This section would not be complete without a reference to man-made forests (plantations) and the impact which they may have on the area of forest land and future wood supply. The importance of plantations intended primarily for wood production is that either they replace forests where net increment is nil or low or are established on non-forest land. A plantation of quick-growing species may, other things being equal, take over the supply function of a very much larger area of virgin (unmanaged) forest.

Under tropical conditions, however, other things are usually not equal: the natural forest generally supplies high-grade tropical hardwoods; the plantation provides quick-grown softwood or hardwood best suited for use as fibre (pulp, paper, wood-based panels) or lower quality sawnwood.

The ecological aspects of man-made forests in the tropics are not at all well understood. To the extent that they might reduce the pressure of exploitation from the remaining natural tropical forest, they are very much to be encouraged. On the other hand, where the natural forest in all its heterogeneity is replaced by monoculture plantations, the ecological dangers are obvious – loss of soil fertility, risk of insect or fungal epidemics, etc. An approach which may be less problematical than regeneration of the natural forest, at least in the tropics and sub-tropics, is afforestation – planting on land which has not recently been forested. Very large areas of such land exist, especially in Africa and Latin America, mainly in the form of grassland (savannah). They are the habitat of wildlife and provide the livelihood for a not insignificant population, partly settled, partly nomadic, and their herds of cattle. Without disturbing either the wildlife or the indigenous populations, however, it is reasonable to suppose that a minor part of such areas could be converted to forestry, with a significant impact on wood supply, as is being achieved in parts of eastern and southern Africa. A positive ecological impact would be the contribution of such plantations to the maintenance of a vegetative cover over an adequate proportion of the world's land mass; also the storage of some of the surplus carbon arising from the burning of fossil fuels and – a more tenuous argument because of lack of supporting evidence – beneficial effects on the macro-climate.

FAO has collected data on industrial plantations in tropical countries and estimated the position in 2000; this information is shown in table 2.6.

In the mid-1970s, there were less than 5 million ha of industrial plantations in the tropics of which two-thirds hardwoods, one-third softwood. By 2000, the area is estimated to exceed 16 million ha. with the most rapid expansion being in the Latin American and Asia-Pacific regions. The production from these plantations, which in the mid-1970s was estimated to be of the order of 10 million m³, could rise ten times to over 100 million m³ and supply about one-third of the total industrial wood removals in tropical countries.

While man-made forests may be the result of either afforestation or reforestation, there is no doubt of the increasing impact which they will have on future world wood supply. No generalization can be made of the relative importance of afforestation and reforestation; it varies from country to country. In the majority of temperate-zone countries reforestation, including the conversion of low-quality woodland or scrub into productive stands, may be the main way in which wood supply can be raised in the long term; in some others, including Spain, the United Kingdom and Ireland, afforestation is of greater significance.

'Natural' forests still account for a significant proportion of the total world-wide, but the tendency is inevitably towards a higher share of managed forests. It could be of crucial importance to the global ecosystem and mankind that this process stops well short of a comprehensively managed world forest, unless management can be advanced to the point where it is as efficient as nature itself in conserving the quality and genetic diversity of the forest ecosystem, especially in the tropical regions. Nevertheless, there is a delicate balance to be struck between the forests' wood-producing role, which involves the need for management on a sustained-yield basis, and its numerous other functions, some of which need management to a greater or lesser extent, others of which depend on the absence of human interference.

Table 2.6 Areas of industrial plantations in tropical countries, 1975 and 2000 (estimated) (million ha)

	Total	Coniferous		Non-coniferous	
		Total	of which: High-yielding[a]	Total	of which High-yielding[a]
1975					
Total	4.71	1.62	0 70	3.09	1 17
Africa	0 91	0.40	0 13	0.51	0.11
Asia-Pacific	1 90	0.26	0.02	1.64	0 27
Latin America	1 90	0.96	0.55	0 94	0 79
2000 (estimated)					
Total	16.39	6 48	2.98	9.91	5.60
Africa	2.09	1.04	0.60	1 05	0 41
Asia-Pacific	6 28	1 96	0.69	4 32	1 38
Latin America	8 02	3 48	1 69	4.54	3.81

Source FAO, State of Food And Agriculture, 1980.

a Generally defined as above a net annual increment of 12-15 m^3/ha/year This implies the principal assortment being produced will be pulpwood.

2.3 World trends in wood consumption and supply

The extent, location and composition of its forest resource is one of the two major factors determining a country's place in the international market for forest products. The other is its demographic, social, economic and industrial structure and stage of development. Regression analysis can determine the degree of correlation between a country's per caput consumption of forest products and a suitable indicator of its economic development, such as its per caput Gross Domestic Product (GDP). Indeed, this correlation has been shown to be significant for most forest products groups and has formed the basis for many of the projections of consumption that have been made at the national and international level.

The closeness of the 'fit' between per caput consumption and per caput GDP tends to vary according to the extent to which other factors, which may form part of the GDP total but which may develop differently from it, have an impact on forest products consumption. The most obvious example is construction, and more particularly dwelling construction, which is in many countries the single most important outlet for the structural types of forest products, sawnwood and wood-based panels. On the other hand, a rather close 'fit' can be observed between per caput consumption of paper and paperboard and per caput GDP, which may be explained by the spread of these products throughout all sectors of the economy.

As an oversimplification, it may be said that the most important wood product of all in terms of volume – fuelwood – has a negative correlation with the level of economic development: the developing countries use on average over three times more fuelwood per person than the developed. However, some developed countries are, for such reasons as the relative availability of wood and other sources of energy, also heavy fuelwood users, such as Finland and the USSR. Fuelwood is among the least significant forest products from the point of view of inter-regional and even of inter-country trade. Nevertheless, its overriding importance to hundreds of millions of people, mostly in the developing regions, assures it a key position in discussions on forest policy, as well as energy policy, even at the international level, as was apparent at the UN Conference in Nairobi in 1981 on New and Renewable Sources of Energy.

2.3.1 Present consumption of forest products

In 1980, world consumption of forest products amounted to 3.44 thousand million m³ EQ (cubic metres equivalent volume of wood in the rough)[6] (table 2.7). This meant that each person used on average about three-quarters of a m³ EQ or almost exactly the same amount as in 1970. In other words, total consumption of forest products grew during that 10-year period at the same rate as population, about 22% or 1.95% per annum. This coincidence may be fortuitous, since population grows at a steady but gradually decelerating rate, while consumption of wood, at least of industrial wood products, follows a more cyclical pattern linked to the fluctuations in the economies of the major consuming countries.

It can be seen in table 2.8 that per caput consumption of the main forest products in each of the developed regions was considerably higher than the world average and, with the exception of Latin Amèrica and Africa, lower in the developing regions. That Latin America and Africa were above the world average was entirely due to their high per caput use of fuelwood, which was more than twice the world average. On the other hand, per caput consumption of industrial wood in the developing countries was on average only one quarter of the world average, while it was three times the world average in the developed regions.

There are some interesting relationships to be observed when comparing the respective shares of the developed and developing regions for different sectors.

	Developed regions[7]	Developing regions[8]
	(Percent share of world total)	
Gross Domestic Product	82	18
Consumption of:		
sawnwood	80	20
wood-based panels	88	12
paper and paperboard	84	16
total industrial forest products	80	20
fuelwood	14	86
total forest products	46	54
Area of forest and woodland	45	55
Area of closed forest	53	47
Standing volume	46	54
coniferous	94	6
non-coniferous	17	83
Population	26	74

The developing regions are predominant in terms of population, non-coniferous forests and production and use of fuelwood. The developed countries are predominant in terms of GDP, coniferous forests and production and consumption of industrial forest products. Because of the developing regions' higher rate of population growth – more than double that of the developed regions – there is no sign that the former are 'catching up' with the industrialized countries in terms of their economies or of their per caput consumption of industrial forest products, although some individual countries have been exceptions.

Apart from population and stage of economic development, other factors which clearly have an influence on the level of consumption of forest products are:

- extent of local wood supply and proximity of forest to markets;
- quality of wood raw material;
- marketing ability of the forest products industries and commerce;
- tradition and culture;
- construction systems;
- climate;
- population density and distribution between rural and urban areas;
- availability and relative cost of competing materials.

These factors help to explain why, in a relatively homogenous region in terms of per caput GDP, such as Europe, there exist considerable differences in consump-

Table 2.7 World consumption forest products in 1970 and 1980 (million units)

	World		Developed		Developing	
	1970	1980	1970	1980	1970	1980
Consumption						
Sawnwood (m³)	414	436	361	350	53	86
Wood-based panels (m³)	70	101	66	89	4	12
Paper & paperboard (m.t.)	128	174	114	146	14	28
Dissolving pulp (m.t.)	4.5	4.4	4.3	4.0	0.2	0.4
Industrial wood used in the round[a] (m³)	140	148	83	74	57	74
Total industrial wood (m³ EQ)	1455	1720	1250	1375	205	345
Fuelwood (m³)	1360	1720	190	230	1170	1490
Total (m³ EQ)	2815	3340	1440	1605	1375	1835
Change 1970 to 1980	*Volume (million units)*	*Percent (%)*	*Volume (million units)*	*Percent (%)*	*Volume million units)*	*Percent (%)*
Sawnwood (m³)	+ 22	+ 5	− 11	− 3	+ 33	+ 62
Wood-based panels (m³)	+ 31	+ 44	+ 23	+ 35	+ 8	+ 200
Paper & paperboard (m.t.)	+ 46	+ 36	+ 32	+ 28	+ 14	+ 100
Dissolving pulp (m.t.)	− 0.1	− 2.0	− 0.3	− 7.0	+ 0.2	+ 100
Industrial wood used in the round[a] (m³)	+ 8	+ 6	− 9	− 11	+ 17	+ 30
Total industrial wood (m³ EQ)	+ 265	+ 18	+ 125	+ 10	+ 140	+ 68
Fuelwood (m³)	+ 360	+ 26	+ 40	+ 21	+ 320	+ 27
Total (m³ EQ)	+ 625	+ 22	+ 165	+ 11	+ 460	+ 33

[a] Pitprops, poles, piling, posts, etc.
Note: Detail may not add to totals, due to rounding.

tion of forest products, e.g. between the Nordic and the Mediterranean countries.

Figures 1 and 2 show why care needs to be taken in interpreting the changes in consumption of forest products between 1970 and 1980, as shown in table 2.7. Compared with most of the previous two decades, consumption trends in the 1970s were far more unsettled, as was the world economy as a whole. It is now widely felt that, at least in the industrialized countries, the phase of dynamic growth in the aftermath of the second world war ended with the first oil-price shock in 1973/74. So far as the developing countries as a whole were concerned, however, the recession of the mid-1970s had relatively little impact on the growth of consumption of industrial wood products. The most recent world recession of 1980/82, on the other hand, seems to have affected both the industrial and developing countries, as may be seen in figure 2.

Table 2.8 Per caput consumption of the main forest products in 1980, by region

	Consumption per caput			Consumption per caput relative to world average		
	Main processed products[a]	Fuel-wood	Total	Main processed products[a]	Fuel-wood	Total
	(m³/EQ[b]/1000 capita)			(Index. World = 100)		
World	329	370	699	100	100	100
Developed	1098	141	1239	334	38	177
North America	1888	107	1995	574	29	285
Europe	840	105	945	255	28	135
USSR	801	293	1094	243	79	157
Japan	1344	6	1350	409	2	193
Australia/New Zealand	1136	80	1216	345	22	174
Developing	79	452	531	24	122	76
Africa	43	904	947	13	244	135
Near East	106	287	393	32	78	56
China	71	157	228	22	42	33
Other Asia/Pacific	54	448	502	16	121	72
Latin America	231	787	1018	70	213	146

Source FAO, Regional tables of production, trade and consumption of forest products, 1970-1980 (FO/MISC/82/16).

[a] Sawnwood, wood-based panels, paper and paperboard.

[b] Cubic metres, equivalent volume of wood in the rough

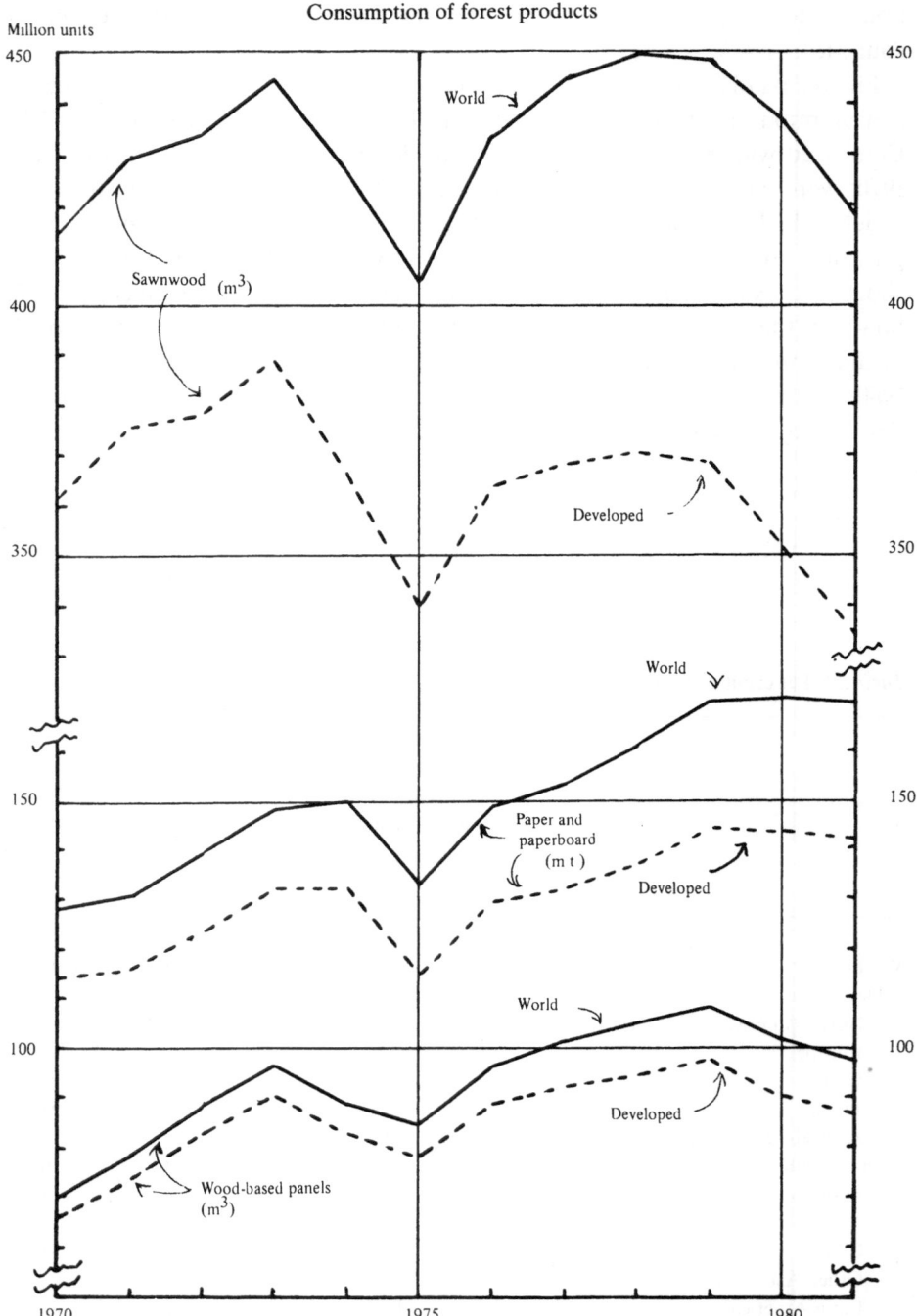

Million units

Consumption of forest products

Figure 1 The 'M-shape' of the trend in consumption of forest products during the 1970s is particularly marked in the case of sawnwood. The fluctuations were much more marked than in the previous two decades.

Trends in consumption of forest products

Legend· SWN = Sawnwood
         WBP = Wood-based panels
         PPD = Paper and paperboard
         DD   = Developed regions
         DG   = Developing regions

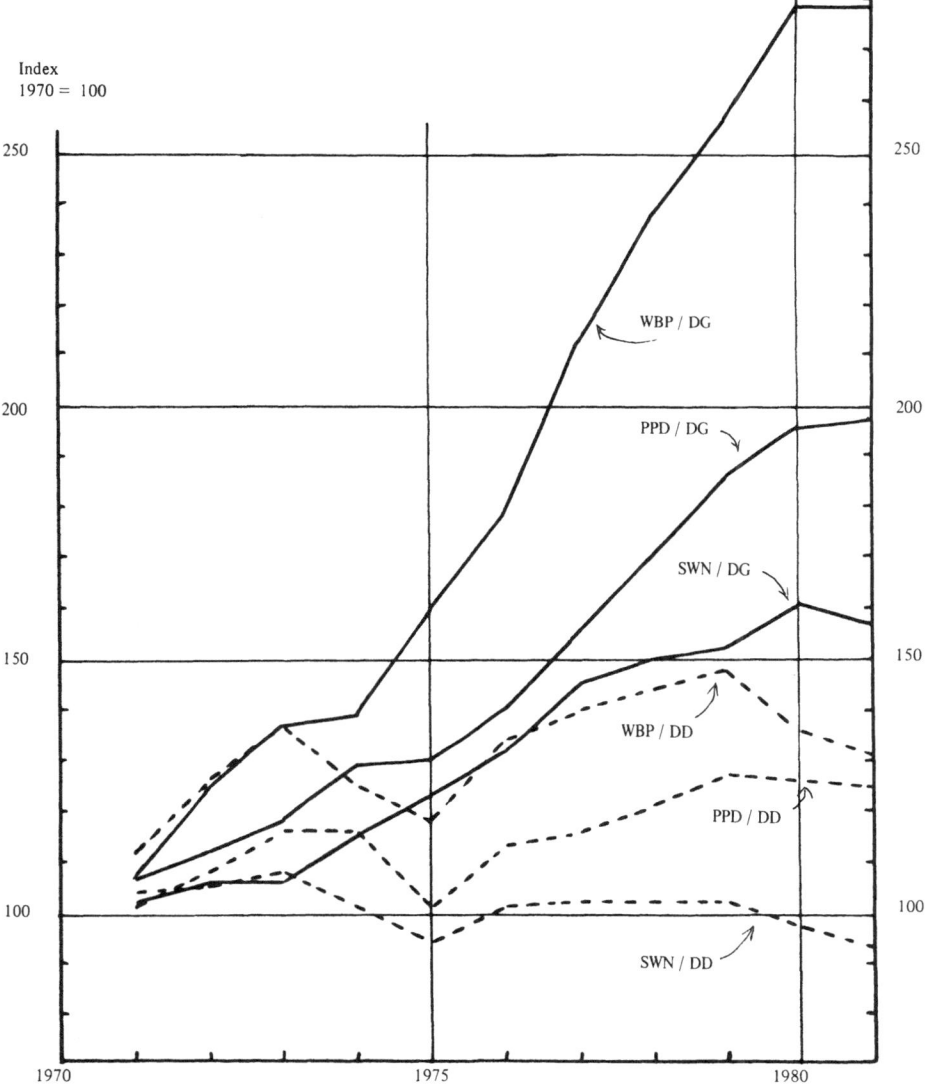

Figure 2. Starting from a very much lower base, in terms of per caput consumption levels, rates of growth of all forest products in the developing regions were far higher during the 1970s than in the developed regions. Growth in the former was also more consistent, although consumption in all regions was adversely affected by the world recession in the early 1980s.

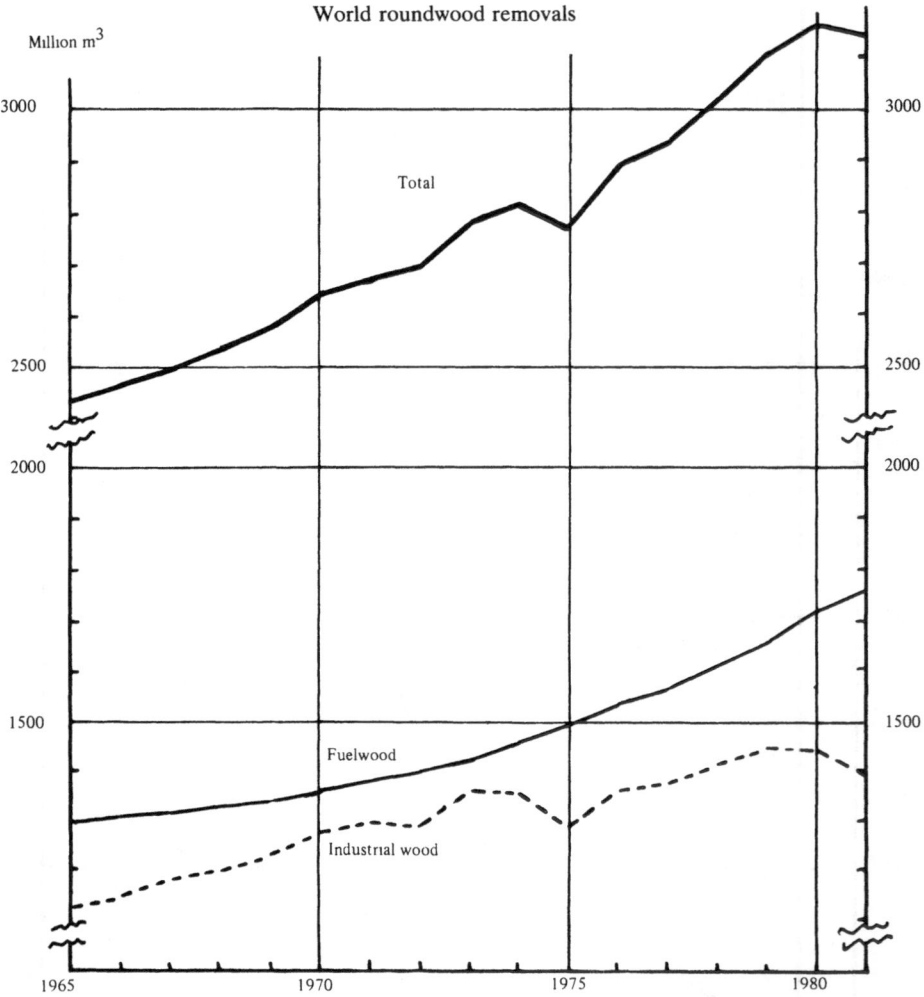

Figure 3. World fuelwood removals have been rising steadily with growth of population; those of industrial wood follow the fluctuations in the world economy. Total world removals rose at an average annual rate of 44 million m³ or 1.6% between 1965 and 1981.

2.3.2 The wood harvest

Countries meet their requirements of forest products from two main sources – their own forests and imports. In the majority of countries, roundwood removals from the forest account for the bulk of total supply to the domestic market.

The volume of world wood removals of 3160 million m³ in 1980 (table 2.9) is only an approximate figure, because, as recent work by FAO and other bodies has shown, considerable differences can and do exist between the officially reported volumes of wood removals and what are probably the real volumes. This is particularly the case for that part of the harvest which does not pass through commercial channels, but is cut and used by the forest owner himself or the local population. The main assortment involved is fuelwood, and it has been estimated that in Europe, fuelwood removals could be 30-40% higher than those reported in official statistics. For several other regions the margin of error, either higher or lower, could be at least as wide, and because of the importance of fuelwood in the developing regions, the volumes involved in such errors could be substantial.

Figures 3 to 5 outline the main trends in removals since the mid-1960s. One of

Table 2.9 Reported removals of roundwood, 1965[a] and 1980

	World		Developed countries		Developing countries	
	1965	1980	1965	1980	1965	1980
Volume of removals	(million m³)					
Total	2422	3159	1211	1348	1211	1811
Coniferous	1008	1220	868	976	140	244
Non-coniferous	1414	1939	343	372	1071	1567
Industrial wood	1131	1441	971	1116	160	325
Fuelwood	1291	1718	239	232	1052	1486
Percent share of total removals	(%)					
Total	100	100	100	100	100	100
Coniferous	42	39	72	72	13	13
Non-coniferous	58	61	28	28	87	87
Industrial wood	47	46	80	83	16	18
Fuelwood	53	54	20	17	84	82
Percent share of world total	(%)					
Total	100	100	50	43	50	57
Coniferous	100	100	86	80	14	20
Non-coniferous	100	100	24	19	76	81
Industrial wood	100	100	86	77	14	23
Fuelwood	100	100	19	14	81	86
Volume change, 1965-1980	(million m³)					
Total		+737		+137		+600
Coniferous		+212		+108		+104
Non-coniferous		+525		+ 29		+496
Industrial wood		+310		+145		+165
Fuelwood		+427		− 7		+434
Percent change, 1965-1980	(%)					
Total		+ 30		+ 11		+ 50
Coniferous		+ 21		+ 12		+ 74
Non-coniferous		+ 37		+ 8		+ 46
Industrial wood		+ 27		+ 15		+103
Fuelwood		+ 33		− 3		+ 41

Source. FAO, Yearbook of Forest Products

[a] The 1965 data for fuelwood were adjusted to take account of later revisions to the long-term series of removals data

the most noteworthy features has been the steadiness of the growth of fuelwood removals (trends over time may be more reliable than actual levels in a given year). In contrast, the growth of industrial wood has become increasingly affected over the past decade by the fluctuations in the world economy. Thus in 1980, fuelwood removals were about 427 million m³ (33%) higher than in 1965; those of industrial wood 310 million m³ (27%) higher. In the same way, the major part of the increase over the 15-year period occurred in the developing countries and consisted of non-coniferous species.

Harvesting intensity is a useful indicator of how fully the resource is being used for wood production. This may be expressed as the relationship between removals and standing volume or between removals and area. The following figures[8] are an example of the former:

Removals as percent of standing volume[9]

	Total	Fuelwood	Industrial wood
World	0.8	0.4	0.4
Developed	0.9	0.1	0.8
North America	1.0	—	1.0
Europe	2.2	0.4	1.8
USSR	0.5	0.1	0.4
Other	1.8	0.2	1.6
Developing	0.7	0.56	0.14
Africa[a]	1.0*	0.9*	0.1*
Asia-Pacific	1.8	1.4	0.4
Latin America[a]	0.3*	0.24*	0.06*

[a] Estimates of standing volume based on out-of-date and incomplete information.

Harvesting intensity is highest in Europe (and Japan, which is included above with 'other developed') and lowest in the USSR and Latin America. The harvesting intensity of industrial wood is very low in the developing regions, with the exception of Asia-Pacific. On the other hand, the harvesting intensity of fuelwood is twice or more the world average in Africa and Asia-Pacific. A high proportion of the fuelwood, however, may come from trees or parts of the tree not covered in the forest inventory.

The harvesting intensity of 2.2% in Europe (equivalent to about 2.5% if bark is taken into account), though more than double the world average, is still less than the net increment of the forest. The standing volume is thus increasing. Elsewhere, where the net increment in mature natural stands is virtually nil, the comparison is more difficult to make. However, the cutting of a natural stand will itself stimulate the rate of increment during the regrowth period. If at the same time, the stand is brought under management, the rate of increment can be further increased. Within limits, therefore, imposed by the inherent or induced fertility of the soil, increment and in consequence removals and harvesting intensity, could be raised without detriment to the standing volume or area of the forest.

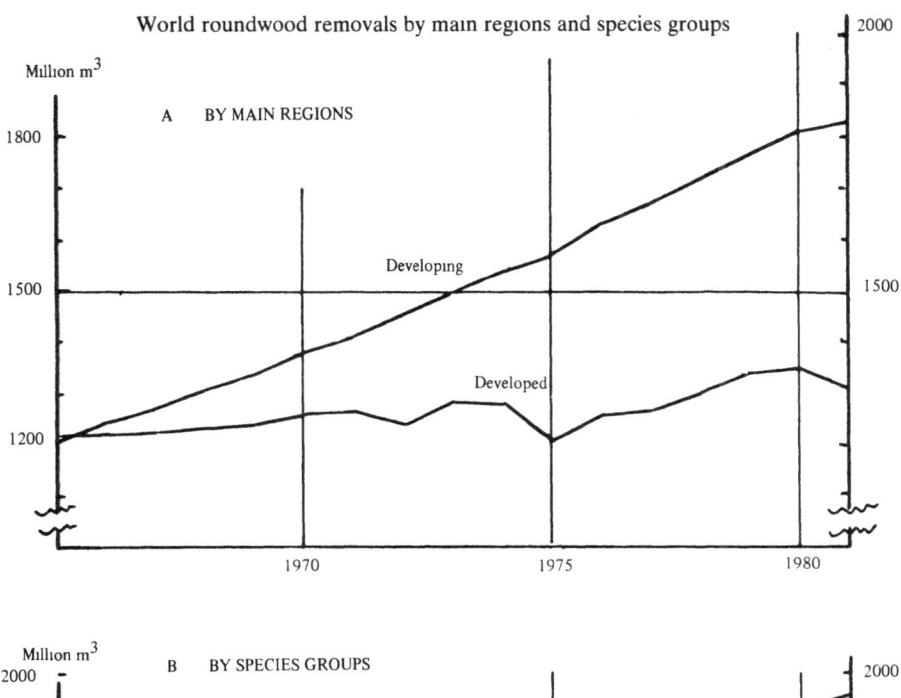

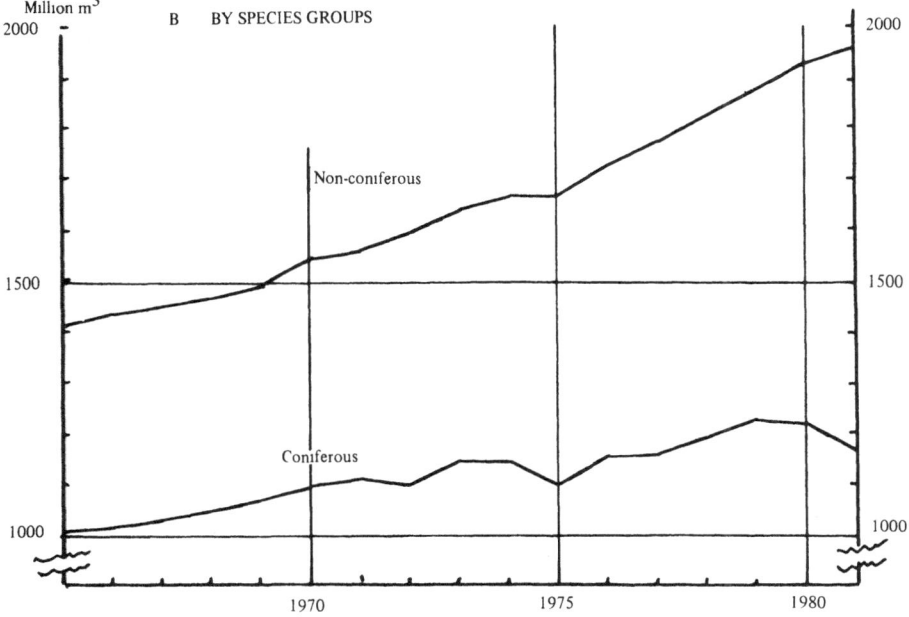

Figure 4. A. The developing regions accounted for most of the increase in roundwood removals between 1965 and 1981 – 617 million m³ or two-thirds out of the world total of 711 million m³.
B. The same trend is apparent for removals of non-coniferous wood. The similarity of trends in removals in developing regions, of non-coniferous wood and of fuel wood (Fig. 3) is striking; similarly, as would be expected, the trend in developed countries, of coniferous wood and of industrial wood. It is more surprising to find, perhaps, that growth in the former was much stronger than in the latter.

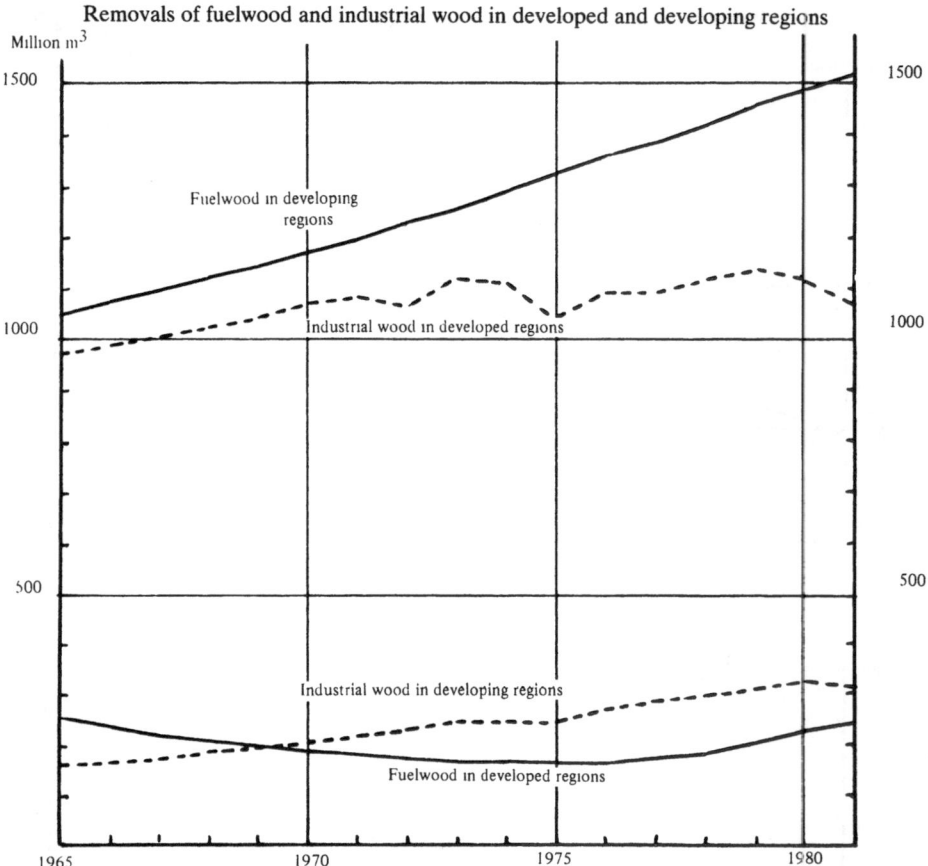

Removals of fuelwood and industrial wood in developed and developing regions

Figure 5. Removals of all categories of roundwood have been rising in the developing regions. Those of industrial wood in the developed regions have shown no clear trend since the early 1970s. The recovery of fuelwood removals in the developed regions since the first oil-price shock of 1973/74 is noteworthy.

2.3.3 *International trade in forest products*

International trade holds the balance between domestic production and consumption. In 1980, as table 2.10. shows, the world value of exports of forest products was about US $ 56 thousand million f.o.b. This puts trade in forest products in third place after energy and agricultural products. Accordingly, for quite a number of countries, foreign trade in forest products is an important consideration in determining forest sector policy, both forestry itself and the forest industries.

Forest industry development, by putting value-added on export earnings, has been one of the cornerstones of policy in countries with substantial forest resources as a contribution to economic and social advancement. A more equitable distribution of the forest industries to take account of the location of the main forested regions has often been discussed at the international level as one component in the strategy to achieve the New International Economic Order called for by the United Nations General Assembly. As may be seen in table 2.11, the developed countries still produced 80% of the world's sawnwood in 1980, 86% of wood-based panels, and 88% of paper and paperboard, while for woodpulp the share was nearly 93%. These countries have less than half the world's standing volume of timber, but on the other hand they do account for the bulk of the consumption of industrial forest products, as described earlier. The debate on shifting the centre of gravity of forest industry should, among other considerations, weigh the merits of location in relation to the main consumption centres as well as to the sources of raw material.

The commodity pattern of forest products trade varies considerably from region to region, and is largely a result of historical factors coupled with existing policies. On the import side, North America imports mainly in the form of processed and semi-processed products; Japan mostly wood raw material. There is a wide range in the import pattern of European countries, the United Kingdom's position corresponding to that of North America, while that of Italy, Spain and Greece is closer to Japan's. But nearly everywhere there has been a gradual trend to reduce the share of raw materials in total trade, fuelled by the desire of exporting countries to increase their export earnings through further processing. This has led to measures by quite a number of countries, both temperate and tropical, to limit or ban the exports of logs. In the case of African and South-East Asian exporters, this is quite understandable given that in the late 1970s more than two-thirds if the volume of exports still consisted of roundwood.

The USSR and eastern Europe are in a somewhat similar situation; in their case about 80% of the export volume consists of logs and sawnwood, with their trade in wood-based panels, pulp and paper still relatively undeveloped. It is noteworthy, however, that several eastern European countries are vigorously developing production and exports of secondary products, notably furniture.

Table 2.12 shows the net trade position for the various regions. There are four major net exporting areas – the three shown in the table, North America, the USSR and Asia/Pacific, to which the Nordic countries and Austria should be added. Their large export surplus is concealed in the even larger net import figure for Europe as a whole. The latter, together with Japan, are the principal net importing areas.

2.3.4 The growing interest in long-term outlook studies

The forest policy debate during the 1970s revolved, more than ever before, around the question of whether the world faced a shortage of wood in the future,

Table 2.10 Value of world trade in forest products in 1980, by product group and region

	Round-wood	Sawn-wood	Wood-based panels	Wood-pulp	Paper and paper-board	Total	Percent of world total			
							Round-wood	Sawnwood and wood-based panels	Woodpulp, paper and paper-board	Total
	(million US $)						(%)			
Imports (c.i.f.)										
World	12394	13895	5150	9687	20450	61576	100.0	100.0	100.0	100.0
Developed	10384	12006	4274	8701	16221	51586	83.8	85.5	82.7	83.8
North America	176	2166	661	1738	3538	8279	1.4	14.8	17.5	13.4
Europe	3381	8238	3422	5601	11107	31749	27.2	61.3	55.5	51.6
USSR	35	63	55	109	582	844	0.3	0.6	2.3	1.4
Japan	6749	1248	74	1074	382	9527	54.5	6.9	4.8	15.5
Other developed	43	291	62	179	612	1187	0.4	1.9	2.6	1.9
Developing	2010	1889	876	986	4229	9990	16.2	14.5	17.3	16.2
Africa	54	198	125	56	487	920	0.4	1.7	1.8	1.5
Near East	144	920	372	42	576	2054	1.2	6.8	2.1	3.3
China	718	34	22	124	337	1235	5.8	0.3	1.5	2.0
Other Asia/Pacific	1049	273	242	405	1298	3267	8.4	2.7	5.6	5.3
Latin America	45	464	115	359	1531	2514	0.4	3.0	6.3	4.1
Exports (f o.b.)										
World	8673	12319	5051	9529	20000	55572	100.0	100.0	100.0	100.0
Developed	4671	10128	3421	8867	19410	46497	53.9	78.0	95.8	83.7
North America	2178	3938	498	4939	5900	17453	25.1	25.6	36.7	31.4
Europe	1427	5060	2593	3325	12209	24614	16.4	44.0	52.6	44.3
USSR	834	980	190	306	363	2673	9.6	6.7	2.3	4.8
Japan	5	38	78	50	709	880	–	0.7	2.6	1.6
Other developed	227	112	62	247	229	877	2.6	1.0	1.6	1.6
Developing	4002	2191	1630	662	590	9075	46.1	22.0	4.2	16.3
Africa	768	163	95	83	25	1134	8.9	1.5	0.4	2.0
Asia/Pacific/Near East	3157	1553	1337	18	289	6354	36.3	16.7	1.0	11.5
Latin America	77	475	198	561	276	1587	0.9	3.8	2.8	2.8

and if so, what should countries do to cope with that situation. The 'shortage psychology' was prompted by the Club of Rome's 'Limits to Growth' which resurrected the Malthusian spectre of an exponentially expanding world population that would come to exert unsustainable pressures on the reserves of land and natural resources, especially the non-renewable ones. The result would be, unless man immediately began to put his house in order, a 'natural' solution to the problem through mass deaths from starvation and pollution, and a break-down in the social structure.

This alarming vision of the future seemed to be supported by the evidence from the effects of the oil-price shock of the mid-1970s and other signs of increasing economic and social disequilibrium, such as rising inflation and unemployment, monetary instability and the apparent inability of governments to find a road back to a reasonably stable growth path.

More recent publications of the Club of Rome have adapted its original model to take account of certain inadequacies, to which atttention had been drawn, notably the fact that insufficient account had been taken of the laws of economics and of the potential of technology to divert part of demand away from resource-intensive to capital-intensive products. An obvious example is the potential to replace fossil fuels by other forms of energy, such as solar and nuclear fusion. Evidence has also been accumulating that man has, either consciously or unconsciously, begun to put his house in order. Population growth is slowing down, especially in the developed countries, but even in many developing countries; economic expansion rates have virtually halved compared with the pre-1973 period; environmental protection measures have slowed down, and in some cases reversed, the spread of pollution; technology is creating a new industrial revolution with electronics; the food problem for a greatly expanded population is still regarded as deeply worrying but potentially solvable, if the necessary political commitment can be mobilised. Last but not least, there has been a growing awareness of the potential of renewable resources.

It is naturally this last aspect which has brought forest resources and forestry into the political limelight more, probably, than at any previous time in history. This interest has been further stimulated by the environmental debate, and a growing public awareness of the multiple-role functions of the forest resource and the dangers of its misuse. As often happens, however, discussions on these vital questions have tended to be conducted more on an emotional rather than a rational plane and have been hampered by an inadequate factual basis. To take an example, even today, despite considerable efforts to clarify the issue, claims as to the rate at which the tropical forest is disappearing range from 5 to 25 million hectares a year.[10] If the true figure is 5 million, the problem is probably containable for the time being, on the supposition that additional land has to be acquired for food production and that losses of natural forest can be offset by afforestation with quick-growing species on a much smaller area. If it is 25 million, this would mean the disappearance of the natural tropical forest al-

Table 2.11 World and regional balances in sawnwood, wood-based panels and paper and paperboard in 1980

	Production		Imports	Exports	Apparent consumption			Self-sufficiency ratio
	Volume	Change since 1970			Total	Per capita	Share of world total	
	(million m³)					(m³ 1000 cap)	(%)	(%)[a]
Sawnwood								
World	439.0	+24.0 (+ 6%)	76.7	79.8	435.9	98	100.0	(100)
Developed	352.4	− 8.6 (− 2%)	66.4	68.7	350.1	300	80.3	101
North America	119.3	+10.5 (+ 10%)	24.5	35.2	108.6	432	24.9	110
Europe	90.5	+ 8.3 (+ 10%)	34.6	25.5	99.6	206	22.9	91
USSR	98.3	−22.2 (− 18%)	0.4	7.2	91.5	345	21.0	107
Japan	37.0	− 5.9 (− 14%)	5.6	0.1	42.5	364	9.7	87
Other	7.3	+ 0.7 (+ 11%)	1.3	0.7	7.9	156	1.8	92
Developing	86.6	+32.6 (+ 60%)	10.3	11.1	85.8	26	19.7	101
Africa	6.4	+ 2.6 (+ 71%)	1.1	0.8	6.7	18	1.5	96
Near East	4.2	+ 1.2 (+ 41%)	3.8	0.1	7.9	37	1.8	53
China	21.2	+ 6.6 (+ 45%)	0.1	0.1	21.2	21	4.9	100
Other Asia/Pacific	29.0	+12.3 (+ 74%)	2.2	7.2	24.0	18	5.5	121
Latin America	25.8	+ 9.9 (+ 62%)	3.1	2.6	26.0	71	6.0	99
Wood-based panels								
World	101.2	+31.6 (+ 45%)	15.4	15.7	100.9	22.7	100.0	(100)
Developed	87.2	+23.9 (+ 38%)	12.9	11.0	89.1	76.2	88.3	98
North America	31.0	+ 4.7 (+ 18%)	2.4	1.8	31.6	125.6	31.3	98
Europe	33.7	+12.0 (+ 55%)	9.9	7.9	35.7	73.7	35.4	94

	(million m.t.)						(m.t./1000 cap)	
USSR	10.6	+ 4 7 (+ 81%)	0.2	1.0	9.8	36.9	9.7	108
Japan	10.3	+ 2 1 (+ 25%)	0.3	0.1	10.5	89.9	10.4	98
Other	1.6	+ 0 4 (+ 33%)	0.1	0.2	1.5	29.6	1.5	107
Developing	14.0	+ 7.7 (+122%)	2.5	4.7	11.8	3.6	11.7	119
Africa	0.9	+ 0.4 (+ 70%)	0.3	0.2	1.0	2.6	1.0	90
Near East	0.8	+ 0.5 (+162%)	0.9	—	1.7	0.8	1.7	47
China	2.1	+ 1.1 (+105%)	0.1	0.9	1.3	1.3	1.3	162
Other Asia/Pacific	5.9	+ 3 0 (+103%)	0.8	3.0	3.7	2.8	3.7	159
Latin America	4.3	+ 2 7 (+160%)	0.4	0.6	4.1	11.1	4.0	105
Paper and paperboard	(million m.t.)						(m.t./1000 cap)	
World	175.1	+47.0 (+ 37%)	33.7	35.1	173.7	39.1	100.0	(100)
Developed	153.4	+35.3 (+ 30%)	26.8	34.2	146.0	124.9	84.0	105
North America	72.8	+15 5 (+ 26%)	8.1	13.7	67.2	267.1	38.7	108
Europe	50.1	+11.3 (+ 29%)	16.2	18.1	48.2	99.5	27.7	104
USSR	8.7	+ 2 0 (+ 30%)	0.9	1.0	8.6	32.4	5.0	101
Japan	18.3	+ 5.4 (+ 41%)	0.6	0.8	18.1	155.0	10.4	101
Other	3.5	+ 1.1 (+ 46%)	1.0	0.6	3.9	76.9	2.2	90
Developing	21.7	+11.7 (+117%)	6.9	0.9	27.7	8.5	16.0	78
Africa	0.3	+ 0 2 (+ 97%)	0.6	—	0.9	2.4	0.5	33
Near East	0.6	+ 0 3 (+ 80%)	1.0	—	1.6	7.6	0.9	38
China	8.8	+ 4.7 (+116%)	0.7	0.2	9.3	9.3	5.4	95
Other Asia/Pacific	4.7	+ 3.0 (+176%)	2.3	0.3	6.7	5 1	3.9	70
Latin America	7.3	+ 3.5 (+ 94%)	2.3	0.4	9.2	24.9	5.3	79

Source FAO Yearbook of Forest Products 1981.
a Production as a percent of apparent consumption.
Note Detail may not add to totals, due to rounding.

Table 2.12 Net trade[a] in forest products in 1970 and 1980 (million units)

	Roundwood (m³)		Sawnwood (m³)		Wood-based panels (m³)		Woodpulp (m.t.)		Paper and paperboard (m.t.)		Total (m³ EQ)	
	1970	1980	1970	1980	1970	1980	1970	1980	1970	1980	1970	1980
World[b]	(+0.8)	(−2.0)	(+3.1)	(+3.1)	(−0.3)	(+0.3)	(+0.3)	(+1.0)	(+0.6)	(+1.4)	(+8.8)	(+13.3)
Developed	−29.2	−28.2	+0.1	+ 2.3	−2.4	−1.9	+1.2	+1.5	+4.4	+7.4	−13.7	+ 3.6
North America	+15.9	+19.8	+5.9	+10.6	−1.8	−0.6	+4.7	+6.8	+3.9	+5.6	+55.0	+84.0
Europe	−16.6	−17.4	−9.5	− 9.2	−1.1	−1.9	−2.6	−4.5	+0.5	+1.9	−44.6	−49.0
USSR	+15.5	+15.3	+8.0	+ 6.9	+0.5	+0.8	+0.2	+0.6	+0.2	+0.1	+32.2	+31.9
Japan	−45.7	−54.0	−2.9	− 5.5	+0.1	−0.2	−0.9	−2.1	+0.4	+0.1	−52.9	−72.1
Other developed	+ 1.7	+ 8.0	−1.4	− 0.5	+0.1	−0.1	−0.1	+0.7	−0.7	−0.3	− 3.4	+ 8.8
Developing	+30.0	+26.2	+3.0	+ 0.8	+2.1	+2.2	−0.9	−0.5	−3.9	−6.0	+22.4	+ 9.7
Africa	+ 6.7	+ 5.8	−0.1	− 0.3	+0.2	−0.1	+0.1	+0.2	−0.4	−0.6	+ 6.0	+ 4.0
Near East	− 0.6	− 0.6	−1.3	− 3.7	−0.1	−0.9	−0.1	−0.1	−0.5	−0.1	− 0.5	−12.7
China	− 1.5	− 7.1	+0.1	− 0.1	+0.6	+0.8	−0.1	−0.3	—	−0.5	− 0.6	− 8.5
Other Asia/ Pacific	+24.9	+26.9	+4.1	+ 5.1	+1.4	+2.2	−0.3	−0.9	−1.2	−1.9	+29.4	+30.3
Latin America	+ 0.4	+ 1.2	+0.2	− 0.2	—	+0.2	−0.6	−0.6	−1.8	−2.0	− 7.3	− 3.4

Source: FAO Yearbook of Forest Products 1981.

[a] + = net exports; − = net imports.

[b] Total world exports should equal imports, but because of various factors, such as differences between countries in classification and conversion factors, timing of shipments and arrivals etc., some differences are to be expected. In 1980 the difference amounted to just less than 3%.

Note: Detail may not add to totals, due to rounding.

together within 50 years with worldwide ecological consequences of catastrophic proportions.

2.3.5 Major long-term studies

The purpose of studies of long-term trends and prospects undertaken by various international and national bodies, the latter both governmental and private, has been to sift and analyse the available evidence about the prospects for the forestry and forest products sector and offer a future scenario or range of scenarios which can be used by policy-makers and planners. Long-range studies are not a particularly new phenomenon. The United States Forest Service has been carrying out timber outlook studies at 10-year intervals since 1909. Since the McSweeney-McNary Forest Research Act of 1928, such studies have been a legal requirement of the Secretary of Agriculture to make and keep current 'a comprehensive survey of the present and prospective requirements for timber and other forest products in the United States, and of timber supplies . . .'. The US Forest Service has remained at the forefront in developing methodologies and models for use in national studies.

It is interesting to observe, however, that even the United States did not, until relatively recently, devote much attention in its long-term studies to international trade aspects, despite being the largest importer and second largest exporter, in volume terms, of forest products.

The first international long-term study was carried out jointly by FAO and ECE and published in 1953. The decision to go ahead with the study was significant in that it marked a budding awareness at the international level that forest reserves were not inexhaustible and that steps might have to be taken to promote wood production through more intensive forest management, if a possible future shortfall was to be avoided. Nevertheless, it took some time for this view to become generally accepted. The findings of ETTS I, the abbreviation for the first European long-term study, were at first criticized for forecasting unrealistically high rates of growth of European consumption of forest products. Events showed, however, that the forecasts for 1960 were too modest and that the main reason for this was the underestimation of the growth rate of Europe's overall economy during the phase of post-war reconstruction.

A valuable by-product of ETTS I was the credibility it brought to long-term studies at the international level. It was followed by studies on other regions, coordinated by FAO. This work culminated in the first effort at a global study. This was presented by FAO to the Sixth World Forestry Congress in 1966 in Madrid (Spain). While this study did not go much further than to bring together the findings of the regional studies into a world scenario, nonetheless it was an important milestone as the first effort to present a global outlook.

During the 1970s, due to changes in policy within FAO towards a more 'action-oriented' programme, its pioneering efforts on long-term studies were not actively followed up, at least at the global and total forest products level. However, it did collaborate with ECE, for example, to produce the third European study (ETTS III), as well as to carry out sector analyses, particularly for pulp and paper and for wood-based panels. Because of FAO's continuing central role in statistical collection and publication, other bodies, which to some extent took over long-term analysis work, depended to a great extent on FAO for basic data.

It became clear during this period that national policy-makers and planners had developed an appetite for supra-national long-term studies. National studies frequently incorporated a review of international trends and prospects, and national decisions on forestry and forest industry developments were being taken on the basis of an international framework.

It was paradoxical that in the latter part of the 1970s, when confusion and uncertainty reached a height as to the future economic outlook, and the economic assumptions underlying the forecasts in some more recent studies already appeared untenable, the wish for some basis for policy-making, however tenuous, should become stronger than ever. Furthermore, the time horizon continued to lengthen. Thus, while ETTS I, published in 1953, made forecasts for 1960, ETTS II (1964) for 1975 and ETTS III (1976) for 2000, interest increased in attempting to align forecasts with the rhythm of forest growth. The earlier studies could be used as basis for industrial planning, but forestry generally needs a far longer time scale to show the results of policy decisions. Under conditions of increasing scarcity and cost of capital, it became no longer acceptable to invite investment in forestry as an act of faith, on the assumption that whatever the future course of economic and technological development, there would always be a need for wood in one form or another; and that anyway the forests had to be maintained for environmental and social reasons.

It was partly with these considerations in mind that the FAO Advisory Committee on Pulp and Paper (ACPP) decided to establish an Industry Working Party (IWP) to undertake an independent assessment of the prospects for the sector. The results were presented to the World Expert Consultation on Demand, Supply and Trade of Pulp and Paper, held in Tunis in 1977. The Consultation recommended that, because of the inter-dependency of different parts of the forest products sector, the IWP's work should be extended to cover the whole sector – in effect a new world timber trends and prospects study.

2.3.6 The world outlook

The study, which involved the mobilization of a large number of experts on a voluntary basis, was prepared, discussed by the ACPP at its annual session in 1979, and published in 1982. It broke new ground in the degree of detail shown – forecasts of demand and supply sub-divided at all levels of disaggregation into softwoods and hardwoods and supply by source (forest, residues by the residue-producing industries, net trade). The basic forecasts are shown in table 2.13.

World demand for industrial wood was estimated by the IWP to rise by 800 million m³ or 62% between 1970 and 2000 to reach 2.1 thousand million m³. Demand for the products of sawlogs and veneer logs would rise somewhat more slowly than that of other industrial wood (pulpwood being the main component of the latter); and for softwood products more slowly than hardwoods. Nevertheless, the products of softwood sawlogs and veneer logs would still be the largest component of total demand for industrial wood, with 41% of the total in 2000 compared with 47% in 1970.

On the supply side, transfer of industrial wood residues was forecast to expand one-and-a-half times to account for nearly 15% of the total (9% in 1970). This would mean removals expanding somewhat more slowly than demand to reach

Table 2.13 World demand and supply of industrial wood in the rough, 1960 to 2000 (forecast)

	1960 (actual)	1970 (actual)	1980 (estimates)	1990 (forecast)	2000 (forecast)
			(million m³)		
Total (softwood & hardwood)					
Demand	978	1291	1470	1742	2085
Products of sawlogs/ veneer logs	647	806	904	1054	1182
Other industrial wood	331	484	566	688	903
Supply (total = demand)					
From the forest	919	1171	1280	1490	1782
Industrial residues	59	120	190	252	303
Softwood					
Demand	721	921	1027	1195	1412
Products of sawlogs/ veneer logs	496	603	675	777	862
Other industrial wood	225	318	352	418	550
Supply (total = demand)					
From the forest	672	823	869	987	1163
Industrial residues	49	98	158	208	249
Hardwood					
Demand	257	370	443	547	673
Products of sawlogs/ veneer logs	151	203	229	277	320
Other industrial wood	106	166	214	270	353
Supply (total = demand)					
From the forest	247	348	411	503	619
Industrial residues	10	22	32	44	54

Source: World Forest Products Demand and Supply 1990 and 2000, FAO.

1800 million m³ in 2000, some 600 million m³ (50%) more than in 1970.

These forecasts were reached after a forcing process to bring demand and supply into balance at the country grouping as well as the world level. This also involved analysis of past and prospective trade balances.

In arriving at the above global figures, the IWP had drawn up forecasts, product by product, and these are summarized in table 2.14 and compared with projections by FAO using its regular econometric approach. Generally the IWP forecasts are below the FAO projections, but only in the case of wood-based panels is the difference striking. This is not the place to discuss the reasons for these differences, which are partly attributable to differences in the underlying assumptions and partly in the methodologies of forecasting.

What is perhaps significant is that, up to the present time at least, all authoritative studies, including those by experts, institutes or organizations[11] other than FAO and the IWP, have predicted a continuing growth in world and regional demand for industrial wood products. Even if this growth is slow, say no more than that of population, this still implies a large volume expansion. Unless or until forest policy-makers receive clear signals of a reversal of this trend, they will have to assume that demand for industrial wood worldwide will continue to expand in the foreseeable future.

Fuelwood, as seen earlier, accounts for almost half the world total of removals and consumption of wood according to the available data, which as noted earlier, can only be taken as approximations. The IWP did not attempt any forecasts,

Table 2.14 Alternative outlooks for the consumption of the main forest products groups

	Actual		FAO projection[a]		IWP forecasts[b]	
	1970	1980	1990	2000	1990	2000
Sawnwood			(million m³)			
World	414	436	503-525	577-626	520	570
Developed	361	350	417-436	463-503	423	448
Developing	53	86	86-89	114-123	97	122
Wood-based panels			(million m³)			
World	70	101	144-194	259-328	141	169
Developed	66	89	126-170	223-279	127	147
Developing	4	12	18- 24	36- 49	14	22
Paper & paperboard			(million m.t.)			
World	128	174	262-286	367-411	256	357
Developed	114	146	215-226	281-304	215	287
Developing	14	28	47- 60	86-107	41	70

Source: FAO, *Agriculture towards 2000* (revised).
[a] Made in 1978, but adjusted to take account of subsequently changed conditions.
[b] Industry Working Party (FAO Advisory Committee on Pulp and Paper).

because it considered them to be outside its competence. FAO has recently undertaken considerable work in this field, however, partly as its contribution to the UN Conference on New and Renewable Sources of Energy, held in Nairobi in September 1980. The problem is concentrated on the developing regions and for them, FAO started with the assumption that per capita consumption of fuelwood might remain constant, that is to say consumption would rise in line with population growth. However, in many wood-poor developing countries, fuelwood supplies are completely inadequate even to maintain current needs; in other developing countries the constraint may be lack of economic availability. FAO concluded, therefore, that future consumption of fuelwood is not likely to increase significantly, except in a few developing countries, mostly in Latin America, and in some will even decline. It estimates fuelwood consumption in developing regions in 2000 at 1700 million m^3, compared with 1450 million m^3 at the end of the 1970s, and a level of 2400 m^3 in 2000 that would be reached if fuelwood consumption were to keep pace with population growth.

For the developed regions, a reversal of the downward trend occurred after the mid-1970s, as was seen in figure 5. While much of the renewed interest in wood as a source of fuel has been from the domestic sector, there has been considerable research and investment in the industrial sector as well. Pulping liquors and industrial residues are being increasingly used for fuel, while the medium- to large-scale use of wood for commercial electricity generation and the production of liquid and gaseous fuels is now technically possible and could become economically attractive, depending on the future course of oil prices. The renewed interest in wood for fuel in developed countries has raised a number of important policy issues: should it or should it not be used as fuel – is it not better used as an industrial raw material? And what are the implications for prices for the smaller assortments, for availability, silviculture, thinning regimes, the economic viability of 'energy plantations' and so on? Debate on these issues is hampered by an inadequate statistical base, as well as uncertainties regarding the future supply pattern and costs of fossil fuels and their possible substitutes.

So far as trade prospects are concerned, there is increasing interest in attempting to incorporate these into demand-supply forecasting. Until fairly recently, regional studies such as the ETTS series for Europe, and national studies, such as those of the US Forest Service, had implicitly assumed that the trade balance was a residual after the internal balance had been established. In other words, it was assumed that if imports had to be found to make up an expected shortfall, they would be forthcoming from other regions; if there would be an export surplus, overseas markets could absorb it.

In more recent analytical work these assumptions are being questioned, and an effort made, despite the difficulties, to assess a given country's or region's trade prospects in the light of the international trade outlook. This was done, for example, in ETTS III (1976) and will be in the next US Forest Service outlook study. Furthermore, a new approach is under way to model the world forest products sector and more specifically to build up a comprehensive picture of international trade in forest products, by IIASA, the International Institute for Applied Systems Analysis. Some results should be available by the mid-1980s.

In today's environment of uncertainty and confusion about the future course of the world's economic and social development, any forecasts, whether of production, consumption or trade, have to be treated with the utmost caution. Unfortunately, this warning is not always heeded by the policy-makers and planners, and too often long-term forecasts are accepted with having more credibility than they merit. Nevertheless, when used in the proper way, the work of forecasters is undoubtedly one of the most useful tools available to policy-makers. Furthermore, it is important that long-term forecasts are kept under regular review, so that those who use them can be kept abreast of latest trends relative to the forecasts, and be provided with early warning of significant deviations from expected trendlines.

2.4 International implications for national policies

No two countries are the same in terms of the extent and composition of their forest resource, or the consumption pattern and trade balance of forest products. Consequently, the extent to which national policy-makers take account of international developments in the forest and forest industry sector will vary markedly. It is particularly through its commercial relations that a country will be directly exposed to external influences; and most countries have in recent decades been expanding the importance of their trade in forest products relative to their production and consumption. Worldwide, the forest products sector has been expanding, more slowly than the overall economy, but not more slowly than population. Studies of long-term trends and prospects have suggested that this expansion will continue in the foreseeable future, although that future has become increasingly difficult to perceive as the number of potentially destabilizing elements in the world economy has grown,

There is no convincingly argued alternative yet to the belief that world consumption of wood and its products will continue to expand, and that the rate of expansion will be higher in those regions and countries where population is still growing strongly and where per caput consumption levels are still low, in other words, the developing countries, than in the rest of the world, the industrialized countries. On the supply side, the developing countries as a group are already worse off in terms of area of forest per inhabitant than the developed countries; and furthermore the forest area in very many of the developing countries is shrinking under the pressure of demand for other land uses, notably food production.

One is drawn inescapably to the conclusion that if the aspirations of the poorer citizens of the world even to maintain their present low levels of wood consumption are to be realized, the pressure on the forest resource in the developing regions will become overwhelming if something is not done to bring the resource under proper management and control, and to raise forest productivity. If it is not, depletion of the forest in those regions will allow wood to be made available for use for some time to come – 30, 50, 100 years? – but sooner or later, the 'renewable resource' will become exhausted. And then what?

Such a doomsday prospect can not be allowed to happen. It represents, however, a challenge to the policy-makers worldwide of a magnitude which too few foresters have yet appreciated. The scale of the problem for the developing world is such that the countries involved have no hope of solving it on their own. It thus becomes the responsibility of the whole international community.

Already, thanks to a number of major developments, forestry has moved into the public and political spotlight in recent times. One development has been the growing awareness, fostered by agencies, such as the World Wildlife Fund (WWF) and the United Nations Environmental Programme (UNEP), of the ecological importance of the forest, and especially of the natural tropical forest. Another has been the growing belief that air pollution may be a major health hazard to forests and trees. In both cases, public concern has preceded a full scientific assessment and understanding of the situation, as well as the willingness of the policy-makers to act. This has underlined two of the basic weaknesses of the position in which foresters worldwide have found themselves: a lack of political 'weight' and an inadequate information base.

Foresters have been well aware of the gaps in their knowledge, but have been hampered by the inadequacy of resources to plug those gaps. There have always been projects of more immediate concern supposedly deserving priority in the allocation of available funds. Nevertheless, the time has come for that misconception to be put right.

The policy-maker has, first of all, to have a sound knowledge of the forest resource in his own country and how it fits into the regional and global picture. This calls for active participation in the work of international agencies, such as FAO, in the collection and dissemination of forest inventory data. It includes the development of new information series to correspond with the growing demand for a wide range of forest functions; and also for better harmonization of classifications and definitions, to ensure international understanding of the information generated.

The same needs are equally apparent for information on the forest industries and the international markets for forest products, even if the quality of data is better, on the whole, than that on the forest resource.

What is needed more than anything else, however, is a clear understanding by the international community of the basic issues which forestry worldwide has to face, and a willingness to provide international support to the national policy-makers who decide to act. Every well-informed person could make up his own list of crucial issues, but those which emerge from the present chapter include the following:

1   *the fuelwood problem in the developing countries, especially in many of the drier areas*. Policy must be addressed to reduce consumption through the use of more efficient stoves and the establishment at the community level of fuelwood plantations. Teaching of the necessary skills and the elements of forest conservation lies at the heart of this policy and, together with the provision of the necessary funding support, provides a great opportunity for international collaboration.

2   *the conservation of as large an area as possible of the remaining natural forest*, with its extremely diverse and possibly vital genetic reserves. This is particularly important and particularly difficult in the tropical regions, and requires a major international effort to support national conservation policies. One especially complex problem is to ensure that such policies are not effected at the expense of the livelihood of the indigenous populations. The building up of a sufficiently strong forest administration to implement the policies is essential and offers scope for international colloboration in training of personnel at all levels.

3   *concentration of management effort on the more productive forest sites*. The range of sites on which a tree crop can be grown is very wide. Given the prospect of capital remaining a scarce factor of production, however, and the fact that growth in consumption of forest products in the developed countries is likely to be modest, it would seem rational to devote resources and effort to raising productivity on those forest sites capable of above-average productivity, especially those well located in relation to the market. In the case of countries largely or partly dependent on imports, a delicate balance has to be struck between raising national self-sufficiency and practising economic logic. Developed countries that are net importers may well find it possible to establish and maintain close trading links with exporting countries – the EEC and the Nordic countries respectively are a case in point. For developing countries, shortage of foreign currency may make the import of forest products a luxury which can be ill-afforded.

4   *the extension of man-made plantations*. Reforestation of felled areas and low productivity woodland are ways in which countries may achieve higher productivity per unit area over the long-term; afforestation of non-forested sites is another. Increment rates achieved on some areas of the tropics and sub-tropics, and even in more temperate zones, have shown that plantations can effectively fill the need for large volumes of fibre. Such plantations can provide the raw material at generally low cost for large capacity industries and may serve a valuable function in meeting needs that otherwise would have had to be met from the natural forest. As yet, experience with man-made plantations is not over a long enough period to ascertain whether there may be disadvantages, such as loss of soil fertility or susceptibility to disease and insect attack, which will partly offset the benefits.

5   *the integration of policies for forestry and forest industries*. That part of national policy directed towards the production of wood must be aligned with industry development. It has happened, in both industrial and developing coun-

tries, that strategies to expand raw material supply have been carried out without adequate consideration being given to whether industries will be attracted to use the material, what those industries will be and whether the wood costs will allow them to compete in the market place. This is where it becomes essential to have a sound background of international information before settling on a national policy. Despite some disturbing signs in recent years of a revival in protectionism, the international forest products market remains a reasonably open one and interdependence has been steadily increasing.

To conclude, the message of this chapter has been that policy making will benefit considerably from a sound appreciation of the international context within which the national forest and forest industry sector is developing. We would go even further and say that, without such an information background and without taking advantage of experiences in other countries, good policy-making at the national level will be difficult, if not impossible. It is evident that the quality and depth of information available about international forestry leaves much to be desired, but whatever its short-comings, it is certainly better than nothing and is improving all the time. Countries have every incentive to co-operate with each other and the international agencies in this task – a case of enlightened self-interest.

If the prospects depicted earlier in the chapter are not too wide of the mark, the developing regions in particular face an increasingly serious wood supply position, which has implications not only for them but also for the industrial countries as well. The problems are not insolvable, especially if public opinion can be mobilised at the international level to support policies for forestry, which are not introspective but face up to the facts that forests are an integral and essential part of the global ecosystem. Equally important, the forests must be managed as a source of a renewable and versatile raw material which will serve man's aspirations for economic advancement. The conservation and careful use of the world's forests is a responsibility that transcends national interests.

Notes

[1] For convenience, the world has been somewhat arbitrarily divided in this analysis into six developed and six developing regions.

[2] Australia, Israel, New Zealand, South Africa.

[3] This section is largely based on 'The International Book of the Forest', Mitchell Beazley, 1981.

[4] Coniferous is synonymous with softwood.

[5] Non-coniferous is synonymous with broadleaved and hardwood.

[6] EQ signifies the volume of wood in the rough which would have to be removed from the forest if no industrial wood residues or waste paper were used in the manufacture of wood products such as paper, paperboard, particle board, fibreboard and dissolving pulp.

[7] For the country separation into 'developed' and 'developing', see table 2.1.

[8] From 'Le marché des produits de la forêt', Peck, Ceneca International Symposium, Paris.

[9] Standing volume overbark, removals, at least of industrial wood, usually recorded underbark.

66

[10] See also section 2.2.4 and the discussion around table 2.5.
[11] E.g. Madas, World Consumption of Wood, Trends and Prognoses (1974); Centre for Agricultural Strategy, Reading, UK, Strategy for the UK Forest Industry, (1980).

References

Beazley, M. (1981) The International Book of the Forest.

Bowman, J.C. et al. (1980) Strategy for the UK Forest Industry. CAS Report 6, Centre for Agricultural Strategy. University of Reading.

FAO (annual) Production Yearbook.

FAO (annual) Yearbook of Forest Products.

FAO (1962) World Forest Inventory.

FAO (1966) Wood: World Trends and Prospects. (Unasylva, Vol. 20 (1-2), Numbers 80-81.

FAO (1978) World Paper and Paperboard Consumption Outlook. Pulp and Paper Advisory Committee.

FAO (1979) Forestry – The 'Jakarta Declaration'. Document for 20th Session of the FAO Conference, C 79/22.

FAO (1979) State of Food and Agriculture 1979.

FAO (1981) Agriculture towards 2000. Chapter 7: Prospects for World Forestry.

FAO (1982) World Forest Products Demand and Supply. Pulp and Paper Advisory Committee, Forestry Paper No. 29.

FAO (1982) Tropical Forest Resources. Forestry Paper No. 30.

FAO (1982) Regional and country tables of production, trade and consumption of forest products. FO/MISC /82/16-20.

FAO/ECE (1953) European Timber Trends and Prospects.

FAO/ECE (1964) European Timber Trends and Prospects, A New Appraisal.

FAO/ECE (1976) European Timber Trends and Prospects, 1950 to 2000.

Madas, A. (1974) World Consumption of Wood – Trends and Prognoses. Akadémiai Kiadó, Budapest.

Meadows, D.L. et al. (1972) The limits to Growth.

Peck, T.J. (1978) Le marché des produits de la forêt. Ceneca International Symposium, Paris.

Perrson, R. (1974) World Forest Resources. Royal College of Forestry, Stockholm. Research Notes No. 17.

3 The production functions

Otto Eckmüllner and András Madas

Wood and other forest products not only furnish many necessities of life but they also constitute the main source of income from forests.

In this chapter, we shall discuss the forest products, the production process in the forest, i.e. forest management, the marketing of wood and the forest industrial implications.

The protection and recreational functions will be discussed in chapter 4, but as already explained in chapter 1, production and these two functions must always be considered together in the broader perspective of forest resource development and as far as practicable pursued in conjunction with one another by multipurpose management in order to achieve optimal resource utilization. Clearly, the weight to be attached respectively to production, protection of the environment and recreation must depend on the site conditions and the owner's objects of management.

3.1 Forest products

3.1.1. Wood in the world economy

Wood is by far the most important forest product overall, although other products such as cork, resin, canes, or even fodder for cattle may be more important locally. That is why the main emphasis in the present context will be on wood. About 90% of all wood consumed in developed countries undergoes much industrial processing before it reaches the consumer as sawn wood, paper or other products, and only about 10% is consumed as fuel. In developing countries, the proportions are reversed with up to 90% of production being used as fuel. Indeed, half the world's population depends on wood to cook food and to keep warm when the weather is cold. Although fuelwood is so vital, the problems associated with the wide variety of industrial wood products are more complex and will be considered first, starting with their share in the structure of world industrial activity as summarized in Table 3.1.

The main points to note in table 3.1 are:

i the first place is taken by the group of metal products and machinery with a very high proportion, both in production and in the number of employees;

ii in second place we find the groups of wood, food and chemical products with shares of the order of 7 to 13%. The wood sector, which includes paper products is thus an important component of the world's economic development. The contribution of wood and forests to human welfare is of course even greater if we take into account the services and indirect benefits which are discussed in chapter 4.

iii Production in all sectors including the wood and paper sector has grown faster than employment; this reflects increased productivity.

It should be noted that since publication of these statistics in 1976, annual growth rates throughout the economy have slowed down. Wood and paper products are no exception, but some growth has continued. Between 1970 and 1980, world consumption of industrial wood rose from 1270 million m^3 to 1390 million m^3 and of fuelwood from 1350 million m^3 to 1630 million m^3 per year.

Table 3.1 The structure and growth rate of world industrial activity

Industrial branch	On the basis of industrial production		On the basis of number of employees	
	Share in 1970 (%)	Average annual growth, 1962-1974 (%)	Share in 1970 (%)	Average annual growth, 1963-1974 (%)
Mining (coal, gas, oil, etc.)	7.4	4.9	5.2	0.8
Manufacturing				
Food, beverages, tobacco	11 1	4.9	12.5	2.0
Wood and paper products	8.6	5 1	10.8	2.4
Chemicals and chemical products	11.5	9.2	7.3	3.1
Other non-metallic mineral products	4.2	6.2	5.3	2.8
Metal products and machinery	41.2	7 5	33.6	2.7
Miscellaneous manufactured articles	10.2	..	22.6	..
Electricity and gas	5.8	7 9	2.7	3.1

Source: Yearbook of Industrial Statistics, UN, New York, 1976.

3.1.2 Factors influencing the consumption of industrial wood

The main factors wich influence a country's consumption of wood are:
- the level of economic development;
- the availability of wood;
- the price of wood;
- the size of population;
- the level of technology.

a   If the level of economic development is expressed by the per capita value of Gross Domestic Product (GDP), the quantity of industrial wood consumed is broadly correlated with it, but the dispersion around the regression line is considerable. A more detailed analysis reveals that:
- the level of per capita wood consumption is higher in the wood exporting countries than in the wood importing countries with comparable GDP levels; this simply means that where wood is plentiful it is used more freely;
- in both wood exporting and importing countries, once a certain level of GDP has been reached, a further increase in GDP will no longer bring about a corresponding increase in wood consumption.

b   The availability of wood depends not only on the total forest area of a country; in some countries, many forests are not in use (e.g. because they are too inaccessible), and so have little influence on industrial wood consumption or production. The forest area in use is therefore more relevant than the total forest area. Other significant factors are the age and species structure and the general quality of the forests in use and of the soils on wich they grow, but it is usually only possible to evaluate the influence of these factors individually in countries with a good system of national forest inventories.

Overall, as already mentioned, for a comparable per capita GDP level, exporting countries consume about twice as much industrial wood as the importers, but the ratio varies for different forest products. For example, taking Northern Europe as an exporting and the EEC as an importing region we find that, whereas at comparable GDP levels the per capita consumption of sawnwood in the EEC attains only 35-40% of that in the Northern countries, the consumption of paper products and of wood based panels is similar in both cases and tends to change in parallel with the GDP level.

c   Economists examining the *influence of long term price trends on wood consumption* generally distinguish between
- 'relative' prices of wood and wood products as compared with the prices of competing materials, and
- 'real'prices where the price index of wood products are expressed as percentages of the indices indicating the general level of wholesale prices.

The 'real' prices of sawnwood rose markedly over a fifty year period: 3 times in the USA (between 1900 and 1950) and 2.2 times in the United Kingdom (between 1901-1905 and 1950). This was one of the main reasons why the amount of sawnwood used, eg per dwelling unit, was gradually reduced.

The 'real' prices of the products of the paper industry on the other hand were fairly stable in the first half of the century, both in the USA and in the United Kingdom. These products are 'new' relative to sawnwood and they are generally produced with more up-to-date equipment and technology.

In the price development following World War II a turning point seems to have occured soon after 1950. Real prices of sawnwood rose steeply during the Korean War and then declined gradually.

Since then sawnwood, after having been ousted from many fields of use, found its way back exactly to those fields where it could compete most successfully; this may have been due in part to the lower price and in part to the technological progress that took place in forestry and wood processing.

The competitive situation of paper products did not change substantially, owing to relatively fast technological progress; therefore their relative and real prices did not change much.

It was after World War II that two new groups of wood-based panel products, fibreboard and particle board, came into extensive use because of their excellent technical qualities. At first they were relatively expensive but, with demand and production increasing, economies of scale and new technical improvements caused their relative and real prices to decline significantly.

Thus, in the postwar period up to the early seventies the price situation of the wood-based products seems to have stabilized in the sense that real prices did not change substantially and some relative prices even declined.

The extraordinary price increases of crude oil from 1973 onwards profoundly disturbed the relatively stable economic growth of a quarter of a century after World War II; it also had a substantial influence on price movements.

One factor affecting prices in Europe and sometimes leading to a certain imbalance between supply and demand is the fact that a large part of the forest resource is in the hands of persons or entities which do not receive their main income from the forests. They can usually wait and not necessarily sell when the industries need the raw-material. Government action under way in a number of countries may remedy this anomaly in the longer term.

Future price levels for wood products remain uncertain but are likely to depend largely, as they have done in the past, on the rate of growth in the world economy; the faster the rate of growth the greater will be the demand for wood and the higher the real price. The fact that a rising demand for wood products has to be met from a world forest resource which is continuing to shrink should eventually push up prices unless these trends are modified by more active forest policies, but perhaps not as rapidly as some forecasters believe. Even if present trends continue, there will be no general shortage of timber until next century. As for fuelwood, there is already an acute shortage in many developing countries and this shortage is continually getting worse.

d *The growth of population* will also have a major influence on future wood consumption. Population growth between 1975 and 2000 has been forecast at 17% for the more developed regions of the world and at 70% for the less developed regions, giving an overall increase of 55% (The Global 2000 Report to the President, 1980). Therefore, even if per capita, consumption remains the same, total consumption will have increased by 55% between 1975 and 2000. A rise in population, however, not only increases the demand for wood but it will also reduce the forest area because more land will have to be cleared for agriculture and settlement.

e *The general progress of technology* affects wood consumption in two ways. On the one hand it leads to the substitution of wood by other materials; for example, the consumption of wood for pitprops in mines, for railway sleepers and building poles has fallen considerably because of the increasing use for these purposes of steel, reinforced concrete and light metals. On the other hand, improved technologies may expand existing uses or create new uses for wood. Timber frame houses and disposable bedding and underwear made from wood pulp are recent examples. It also happens frequently that an improved wood product ousts a traditional one. Thus, as a packaging material, wood-based panels, paper and paper board are being increasingly substituted for sawnwood and this leads to a substantial saving of wood because, in order to package a given quantity of a commodity, more than twice as much wood is needed if the packaging is made of sawnwood than if it is made of fibre board or plywood. This saving in wood makes the new wood-based product relatively less expensive than competing non-wood products and is therefore likely to stimulate the total demand for the wood product. The indications are that technological progress in the industries producing competing materials will continue to gather momentum. If wood is to retain its competitive position, there must be equal progress in the wood products sector. This is an important point which those responsible for forest policy must not overlook. Wood products research is receiving far too little attention in many countries of the world. An important line of development is the use of wood in combination with other materials, eg. in packaging.

3.1.3 The industrial wood products

a *Sawnwood:* Sawnwood is the traditional forest product: its share of industrial wood consumption was at the beginning of this century decisive. Since then, it has gradually been losing its predominant role. Technological progress in the use of competitive materials, such as steel and concrete on the one hand, and the favourable properties of new wood-based products, such as panels on the other hand, are responsible for this gradual decline in sawnwood's share of total industrial wood consumption. In Europe, this has developed as follows:

Period or year	%
1913	59
1935-38	56
1961-71	44
1980	44

Broadly speaking, the trends in developed countries in other parts of the world have been similar.

World consumption of sawnwood (including sleepers) has remained static at a little over 400 million m^3 per year since 1970 and per capita consumption has decreased from 112 m^3/1000 capita to 92m^3. The increase in population during the period has thus compensated for the reduction in per capita consumption. In Europe, consumption has remained at around 90 million m^3 per year and consumption per 1000 capita has remained static at around 200 m^3 per year, but with considerable annual fluctuations caused by changes in general economic conditions. Looking back further annual consumption per 1000 capita in Europe has hardly changed since the beginning of the century; in the USA however, there was a sharp decline during the first half of the century and in the USSR a steep rise; in both these heavily forested countries, the annual consumption per 1000 capita has now stabilized at around 450-500 m^3 ie at rather more than twice the European figure. In Japan annual consumption has become stabilized at around 300 m^3 per 1000 capita. More than half of all sawnwood is used for construction; another 15-20% goes into packaging; a wide variety of uses including furniture, mining, railway sleepers account for the rest. Consumption for the latter two uses has been declining steadily and only remains significant in a few countries.

b *Wood-based panels. Plywood:* Veneer plywood is plywood manufactured by bonding together more than two veneer sheets. The grain of alternative veneer sheets is crossed, generally at right angles. Core plywood is plywood whose core is solid and consists of narrow boards, blocks or strips of wood placed side by side which may or may not be glued together. Cellular board is a plywood with core or certain layers made of material other than solid wood or veneers.

Particle board: A sheet material manufactured from small pieces of wood or other ligno-cellulosic materials (e.g. chips, flakes, etc.) agglomerated by use of organic binder together with one or more of the following agents: heat, pressure, humidity, a catalyst, etc.

Fibreboard: A panel manufactured from fibres of wood or other ligno-cellulosic materials with the primary bond deriving from the felting of the fibres and their inherent properties. Bonding materials and/or additives may be added.

Though veneer and plywood are long-standing, classical products, the development of consumption of wood-based panels as a whole became more significant only after World War II, when two entirely new products, fibreboard and particle board started to be used on a mass scale.

In contrast to sawnwood, the aggregate consumption of wood-based panels has been rising at a rate exceeding general economic growth. To a great extent they have been replacing sawnwood, because of their relatively lower prices and excellent technical properties, eg. constant size and quality, greater strength, larger choice in width and length, a well finished surface, and consequently smaller amount of labour required for their fitting, mounting and finishing.

Moreover, most panels can be produced from low quality coniferous or broad-leaved roundwood of small dimensions, unsuitable for making sawnwood, and even from wood residues; panels may, therefore, replace sawnwood and plywood made of higher quality, expensive logs. This clearly has implications for forest policy, but it would probably be wrong to conclude that the markets for high quality logs will diminish and that there is no longer a need for foresters to continue to grow trees that will produce such high quality logs. The demand for luxury timbers is in fact likely to remain firm, especially as it has to be met from a diminishing resource of virgin forests.

World consumption of panel products rose from 66 million m^3 in 1970 to 96 million m^3 in 1980. The most significant increase was in particle board, the consumption of which more than doubled from 19 to 42 million m^3. More details are given in table 3.2, which shows the breakdown by products, and the corresponding figures for Europe as an illustration of the fact that the relative importance of the various panel products is by no means similar throughout the world. Thus, by world standards, Europe is a very high consumer of particle board. A point that is not apparent from the table is that total annual consumption of all panel products, including particle board, levelled off or even dropped slightly towards the end of the 1970's and the beginning of the 1980's. The main cause is the world economic recession during that period. The future, as already mentioned, is likely to lie with improved or new panel products. Promising examples are medium density fibre board and lamin boards.

c *Paper and paper board.* These are products made mainly of wood pulp although other materials, such as bagasse are also used where they are available. Wood pulp is defined as fibrous material prepared from wood by mechanical and/

Table 3.2 Consumption of plywood, particle board and fibre board (million m^3)

| | World | | Europe | |
	1970	1980	1970	1980
Plywood	33	38	4.8	5.1
Particle board	19	42	12 3	24 1
Fibre board	14	16	4 2	4.6
Total	66	96	21 3	33.8

or chemical processes for further manufacture into paper, paperboard, or other cellulose products.

Paper and paperboard have become, during the past decades, the most important industrial forest product. The world consumption between 1970 and 1980 developed as follows:

	World Consuption (million tonnes)	
	1970	1980
Newsprint	22	27
Printing and writing	27	42
Other paper and paper board	79	105
Total	128	174

About 75% of the category 'other paper and paper board' are wrapping and packaging materials and the remaining 25% are divided more or less equally between household and sanitary paper on the one hand and a large variety of speciality products on the other hand; they range from photographic sensitizing paper and filter papers to gasket boards, suitcase board and wall papers.

World consumption of paper and paper board in 1980 was distributed as follows:

Country	%
USA	36
Europe	28
Japan	10
Other developed countries	9
Developing countries	17
Total	100

The general recession in the world economy during the late 1970's and early 1980's is probably responsible for the fact that total consumption has stagnated and that there has even been a slight drop in the per capita consumption of paper and paper board.

Looking to the future, in the developed countries, a continuing rapid increase in the consumption of household and sanitary paper seems probable, e.g. for disposable clothing and linen substitutes for use in hospitals and hotels as well as in the home. The advent of the 'electronic office' and electronics in the home while generating paper consumption for print outs, could on the other hand reduce the demand for newsprint and printing and writing papers, but this is still doubtful. The demand fot these last two categories is likely to rise rapidly in developing countries with the spread of literacy.

d   *Other industrial roundwood.* Most wood in this group is used in the round for mining (pit props), communications (poles), construction (scaffolding and framework support), and agriculture (cheap farm buildings, fences etc.), but the FAO statistics fot this category also include roundwood used for tanning, match blocks, gazogenes, distillation etc. World consumption of wood for these purposes remained more or less constant at just over 200 million m^3 per year between 1970 and 1980, but with a modest decrease in the developed countries and a corresponding increase in the developing countries. The USSR alone accounts for almost half of world consumption. The use of roundwood as a chemical feedstock will be discussed under end uses.

e   *Fuelwood.* World consumption of fuelwood at 1630 million m^3 in 1980 exceeded that of industrial wood which was 1390 million m^3 and continues to rise somewhat faster. As already stated, 90% is consumed in developing countries where most people in rural areas have no other fuel for cooking and heating, a situation which is not likely to change for generations. Even in oil rich developing countries, rural populations will continue to depend on wood, because to bring oil, gas or electricity within their reach would require prohibitively high capital expenditure on roads and on electric transmission lines. In towns where populations are more concentrated the position is of course different although also here much wood is used in some countries, especially in the form of charcoal which is easier to transport than wood and more convenient and cleaner in use. On the other hand, about one third of the heating value of wood is lost by conversion to charcoal. World production of charcoal is 16 million tonnes per year, and is rising steadily. Shortage of fuelwood has been called the poor man's energy crises; it might be even more to the point to call it the poor woman's energy crisis because it is usually the women who have to find the fuelwood and in some instances carry it for long distaces. These problems are discussed further in chapter 5.

Mainly in some developing countries, fuelwood continues to be indispensible for a variety of local industries and crafts: curing tobacco, baking bread, burning bricks, drying tea, making pottery and brewing beer, to name a few.

In the developed countries, there was a fairly rapid decline in the use of wood for fuel up to the mid–1970's because even farmers with their own forests turned to other more convenient and relatively inexpensive fuels such as oil, gas or electricity. The drastic price escalation of oil, however, has changed the position. Especially forest owners, but also others, are again using more fuelwood for heating. Overall this increase may not be great and is unlikely to be reflected fully in published statistics, because it is often the very small woodland owner who consumes what he himself has produced.

In most countries, the increased use of wood for fuel has not competed with industrial uses because the wood is either of a quality too poor, or the supplies are too scattered or distant from wood–processing industries for economic marketing as industrial raw material. There are, however, exceptions, and in Sweden for example, there has been some concern that pulp mills might obtain less wood because of the competition from fuelwood.

The interest in the use of fuelwood for other than domestic purposes has also revived. In the wood processing industries themselves, increasing attention is paid to the generation of energy from wood residues not needed for further industrial processing (sawdust, bark, 'black liquor' etc). Plans to use wood to generate electricity have been drawn up or are under consideration in various parts of the world, for example, in the Irish Republic and in the Philippines. In some small towns and villages in Western Europe, district heating systems have recently been installed. Just where these developments will lead is at present difficult to say, but they all depend on a sufficient wood supply within very short and inexpensive transport distance from the point of consumption.

Charcoal, at one time, was used on a large scale for smelting iron and other metals, but this has ceased with few major exceptions such as Brazil where considerable plantations are created for that specific purpose. Very high quality charcoal is still used on a small scale in the chemical industry for a variety of purposes, for example to act as a filter.

3.1.4 Trends in the structure of wood consumption

In the previous section, a general picture was presented of the structure and development of wood consumption by the major forest industries. This section is intended to show the changing consumption patterns, to identify the major end uses of forest products and to discuss some of the economic and technical factors which influence the consumption of forest products in particular end use sectors.

The data used in this section are taken partly from the end use sector studies

Table 3.3 Consumption of industrial wood in Europe by major end-uses

End-uses	Wood raw material equivalent (million m³/WRME)			As a percentage of total Consumption (%)		
	1950	1960	1975	1950	1960	1975
Construction	49	70	140	29	29	38
Packaging	26	45	88	15	18	23
Furniture	11	20	29	6	8	8
Mining	19	20	10	11	8	3
Railway sleepers	6	5	4	3	2	1
Printing and writing paper	12	23	47	7	9	13
Textiles	4	7	14	2	3	4
Other	47	56	40	27	23	10
Total	174	246	372	100	100	100

Source· ETTS I, II, III.

and partly from the European timber trend studies prepared for the ECE Timber Committee and the FAO European Forestry Commission in recent years.

For comparison of the different wood products, in such calculations, the physical volumes of the products are converted to 'wood raw material equivalent', or WRME, meaning the amount in cubic metres of wood raw material needed to produce a cubic metre of sawnwood, a tonne of paper, etc. In many cases, European totals were estimated on the basis of data from a limited number of countries. For other reasons as well, a high degree of accuracy cannot be claimed for the figures, although it does appear that the overall picture is realistic and offers a sound basis for those who are dealing with forest policy and want to know where the products of forestry and forest industries are mainly used.

Longer term trends in the structure of consumption of wood products by end uses are best illustrated by Europe for which statistics go back farther than for other parts of the world. A summary is given in table 3.3.

The table demonstrates that the main growth sectors have been construction, packaging and printing and writing while the main sectors to suffer decline have been mining and the category 'other' which, as previously mentioned, comprises a variety of uses, mainly of wood in the round.

Unfortunately, there are no exactly comparable figures from other parts of the world, but the available evidence suggests that construction takes an even bigger share of total consumption in North America and some other countries, than in much of Europe. The various end uses will now be considered in turn.

a *Construction.* We have seen that sawnwood, panel products, wood used in the round and even paper products all play a part in construction which has not only maintained but even consolidated its position as the main end use of industrial wood, accounting for more than a quarter of all industrial wood consumed and well over half of sawnwood and wood-based panels. A large proportion of dwellings in the developed countries in the forest rich regions of the Northern hemisphere are mainly built of wood. This is the situation in Canada, Finland, Japan, Norway, Sweden, the USA and USSR. In Japan, around 90% of dwelling houses are of wood, in Norway some 60%. Wooden houses have also started to spread, as a result of promotion campaigns, in countries with limited forest resources, such as the United Kingdom and the Irish Republic.

Much wood is used in agricultural buildings. In some countries of Eastern Europe, for example in Hungary, the consumption of wood and wood products in buildings used for animal husbandry has increased considerably since the early 1970s.

The main reason is that new technical processes have been introduced and adapted for transforming wood into elements of prefabricated buildings, as a result of which the woodworking industry is now able to use wood species and dimensions, which were up to now unsuitable for construction. The innovations made wood products more competitive in construction in terms of prices as well as technological characteristics (productivity by prefabrication, long glued and bent beams, resistance against chemical vapours etc.). Another field where wood is gaining ground in construction is the building of weekend houses, which is connected with the increase in living standards and leisure time. This is occurring even in regions where wood is generally playing a minor role in house building.

Among the many factors which will determine the future consumption of forest products in construction, four may be mentioned as important:

- techniques for prefabricated building construction are likely to grow in importance;
- stress grading, finger-jointing, glueing, bending and other technical innovations are making sawnwood into an engineering material whose properties are known precisely. This should encourage the rational and economic use of sawnwood and encourage architects and designers to specify it with full knowledge of its characteristics and confidence in its performance;
- new products are appearing which are pushing forward the application of wood in construction. Such a product is the cement-bonded panel, which is better from the point of view of fire safety than other building materials such as steel;
- building regulations which control the use of wood products in some countries are out of date and may sometimes not reflect technical advances. Fire safety regulations are a common example. If, through pressure from specialists and publicity, building regulations are updated, we can expect a bright future for wood products in construction.

b *Packaging.* In packaging, as in construction, sawnwood, panel products and paper all play a part. Packaging has been increasing and its share of industrial wood lies between 10% and 25% in most developed countries. In terms of value about half of all packaging is wood-based. Metals, glass, textiles, plastics and other materials account for the other half. These proportions do not seem to be greatly affected by the level of economic development, but there is a trend for the share of sawnwood and panels to decrease and that of paper and paper board to increase. Without the wood-based packaging materials the transport, handling and display at the point of sale of many goods would be almost impossible and trade would be badly hampered.

c *Furniture.* The level of furniture purchases is closely related to the level of construction of new dwellings; but fashion also plays a part.

Before World War II furniture was generally a 'once-in-a-lifetime' purchase. This practice changed after World War II. Higher living standards have given many more people than before the opportunity to establish families at a younger age, to have a home of their own and to furnish it. The consequence of this has been the growing importance of mass production in the furniture industry and the decrease of average length of life of the furniture, which became subject to similar changes in fashion as other consumer durables.

Following general trends, with the further rise in living standards, more and more families are turning back to furniture made of 'real wood'; this is one of the reasons why in recent years expensive species, such as oak and teak, have received particular attention.

According to ETTS III the main trend in the raw material input of the furniture industry was the growing substitution of sawnwood by wood-based panels, which are well suited to modern mechanized manufacturing techniques, and especially by particle board, which has been improving its price advantage. Metal and plastics are also increasing their share of the raw material input of furniture. All these substitution trends are expected to continue. In Europe, the share of wood (sawnwood and wood-based panels) in the raw material consumption of the furniture industry has been estimated to have changed between 1969 and 1980 as follows:

	1969	1980
	%	
Chairs	76	67
of which legs and frames	61	50
Upholstered furniture legs and frames	95	86
Table legs	83	64
Drawers	98	80

The share of metal and plastics increased correspondingly. This proportional decline in the use of forest products was offset by the expansion in the furniture industry as a whole.

d *Paper for printing and writing.* The consumption of printing and writing paper has grown strongly since the 1950's; it doubled its share of industrial wood consumption in the last quarter of a century.

Printing and writing paper is used over an enormous field of human activity from newspapers to computer printouts and bank notes. The two main sectors are newsprint and other printing and writing paper.

– *Newsprint.* The tonnage of newsprint consumed is governed by the number of newspapers printed, the number and format of pages in each copy and the thichness of the paper. There is one remarkable difference between the newspapers of Western countries and those of the Eastern European ones, the role of advertising in the newspapers. In the West, the amount of advertising in newspapers plays an important role both financially and for the tonnage of newsprint consumed. Newspapers are holding their own, even though television is now firmly established. A long term co-existence between the different forms of news media now appears probable.

In the Eastern European countries, advertising is unimportant and the size of newspapers is not linked directly to economic activity. Recently however, advertising seems to be gaining ground in the newspapers in some countries in Eastern Europe, partly for economic reasons, partly because the gradually improving economic regulators are forcing the enterprises to pay more attention to market conditions and to sell their products. The ratio of newsprint consumption to population (Kg/caput) has developed as follows:

	1950	1980
Nordic countries	12	30
European Economic Community	6	17
Central Europe	6	20
Southern Europe	1	5
Eastern Europe	2	7
USA		46
Japan		23
Europe	5	13

Source. FAO

Other printing and writing paper. The consumption of other printing and writing paper has grown also strongly since the 1950's for many reasons, the most important of which are the following:

- Illiteracy is now almost entirely overcome in Europe, so that virtually everyone from school age upwards is a potential user of printing and writing paper;
- General progress of technology has created new uses for paper, for example, in computer systems;
- On the one hand the management of the economy has become more complex, and on the other hand people have more leisure time and more disposable income. These two factors are also promoting the consumption of printing and writing paper for books, magazines, brochures, correspondence, office forms, etc.

The extrapolation of past trends would suggest a further increase in the consumption of printing and writing paper. But the situation in the field of communications shows such potential for radical change that any forecasts of the consumption of printing and writing paper must be treated with special caution.

e   *Mining.* There has been a spectacular decrease in the consumption of mining timber both sawn and round (pitprops); advances in mining technology and competition from steel (sometimes used in conjunction with wood) are among the main reasons. Mining timber now amounts to less than 3% of world consumption of industrial wood and a substantial increase seems unlikely even if coal mining should again expand. There are, however, a few countries where mining timber continues to be of some significance. China and the USSR between them account for almost two thirds of world consumption.

f *Railway sleepers*. The consumption of wood for railway sleepers has also been in decline for many years partly because wooden sleepers have been replaced by reinforced concrete and steel and partly because, in some developed countries, railway systems, far from continuing to expand, have begun to shrink. The share of sleepers in the world consumption of industrial wood is now less than one per cent.

g *Wood used in the chemical industry*. The production of dissolving pulp for the manufacture of cellulosic fibres such as rayon has been the main use of wood in the chemical industry (excluding chemical pulp for paper). World consumption of dissolving pulp has been around 4.5 to 5 million tonnes per year since 1970 or about 15% of the total consumption of textile fibres of 26 million tonnes. A study prepared for the ECE Chemical Industry Committee expects that world requirements of textile fibres will double within twenty-five years, but those of cellulosic fibres are expected to remain stable, so that their share of the total would decline to around 7%. The shares of cotton and wood are also expected to decline but synthetic, oil-based fibres are expected to continue to expand their share to around 60% of world textile requirements.

The technology for manufacturing a wide range of other chemical products from wood and wood residues has already been developed; they range from fuels such as methane, methanol and ethanol to polysaccarides, epoxyresins, polyurethane resins, formic acid, vanillin and sugars. Goldstein (1974) goes as far as to claim that 'cell wall polymers which constitute the major portion of wood have the potential for meeting all of our chemical needs in place of petrochemicals.'

At present, however, it is generally still more economic to manufacture these chemical products from coal and oil and most experts believe that this situation will continue until fossil fuels become less abundant. That is why these wood-based chemical industries are mostly in their infancy. There are, however, exceptions. In Brazil for example, methanol made from wood is used on a commerical scale to complement petrol as a fuel for motor vehicles. Methanol is also an intermediate stage in the manufacture of various chemicals from wood.

3.1.5 *Forest products other than wood*

Some of the products other than wood are derived from the trees, some from other plants and from animals occurring in forests. Some are traded commercially, others are traded little if at all and therefore tend to be ignored in statistics and by foresters even if local populations would suffer hardship if they were deprived of these products.

Among the traditional commercial products of importance in several countries are resins, gums, cork, and bark for tanning; these products are now facing increasing competition from substitutes. Other important forest products like rubber are now mainly produced in special plantations outside the forest. In some developed countries the production of Christmas trees and ornamental tree foliage, originally a minor adjunct to forestry, is developing into a separate multi-million dollar industry. Bark is acquiring greater commercial significance for horticultural purposes as a mulch or compost and it is also burnt in wood using industries to generate energy.

The products other than wood, especially those of little commercial value, are generally far more important in developing countries than in developed countries partly because the people are more dependent on them and partly because there is a much greater variety since many developing countries are in the moist tropics where the vegetation is particularly luxuriant.

In developed countries people may enjoy picking edible berries or fungi, but few depend on these activities for a living. Wildlife management to provide the necessary conditions for hunting is a significant income generating production function of forestry in some countries. The hunting itself is a form of recreation which is dealt with under that heading in chapter 4. Forest grazing and litter for cattle used to be important for European agriculture and caused serious forest damage, but wherever agriculture has advanced sufficiently, these demands on the forest have virtually disappeared, or are met by modern systems of agroforestry which permit the production of timber and animal feed on the same piece of land without harm to either interest. In a few specific cases products such as resin or cork may take precedence over timber in forest management; but generally speaking non wood products no longer are of major importance.

In developing countries, the position is very different. This is a point sometimes overlooked by foresters from developed countries when they first start working in developing countries where many people depend on forests for animal feed, fibres, food for human consumption and medicines. Poulron (1983) describes this in some detail for Africa. Feed from forest vegetation may benefit two groups of animals, wildlife and livestock which in turn yield products satisfying human needs such as meat, milk, hides and eggs. The meat obtained by hunting, in contrast to developed countries, may be an essential source of protein. Forest fibres are used for making baskets, mats, rope and furniture; they are also used in house construction. The sources include various leaf fibres, the outer bark of climbing palms, palm fronds and the bark of certain trees. Forest foods include foliage, fruits, gum (where consumed as food), honey, insects, meat, oleaginous products, vegetables from the forest floor and fungi. Drugs and medicines have been obtained from the forest since times immemorial. Some have found their way into modern medicine; most have not yet been tested scientifically. The enormous variety of useful species is illustrated by a publication issued by the Subsecretaria Forestal y de la Fauna (1981) of Mexico; this lists 200 species of forest plants and 48 species of animals which contribute directly to human diet in Mexico, and 70 species of forest plants which provide animal feed in that country.

Excessive or irrational exploitation of these resources which are often taken for granted until they are no longer there, has often led to their exhaustion and even to complete forest destruction. Uncontrolled grazing, especially by goats and shifting cultivation where population pressures are high cause the worst damage. The policy implications of these problems are discussed in chapter 5. Any major change in forest management such as the conversion of mixed natural or semi-natural forest to plantations of a single species may also profoundly influence the supply of some of these products. Policy decisions concerning such changes should, therefore, not be taken without considering the needs of local populations for these products. There is also the broader issue of preventing the extinction of animal and plant species which may depend on a particular habitat for their survival and whose potential value may not yet have been explored.

3.2 Forest production

3.2.1 Introduction

The production problems facing world forestry may be summed up as follows: By the year 2000 the world will need an extra 500 million m^3 per year of industrial wood and a similar increase in the production of fuelwood. To achieve these increases we require from now on an annual investment of the order of 5 billion dollars for forestry and about twice that amount for new industries. Existing policies cover only about 20% of these requirements.

The expression 'forest production' has two distinct meanings: foresters understand by forest production the biological production, that is the annual growth of a forest, the increment, as a rule stated in cubic metres per hectare (m^3/ha), normally overbark, i.e. including the bark. Timber industrialists or salesmen take forest production as the *technical production*, that is the wood harvested and removed annually from the forest (mainly measured in m^3 underbark), which is delivered to the market or used by the forest owners themselves.

However, as already discussed earlier in this chapter, forests do not produce wood only and in some regions other forest produce can be of considerable importance both for domestic consumption and commercially.

Forests belong to the very few renewable natural resources, but they need to be properly managed, not over-exploited or mis-used if they are not to be exhausted or destroyed. The demand for wood will in future increase most probably faster than in the past and to produce more and better wood will become increasingly necessary and urgent.

3.2.2 Past developments

In some parts of the world a growing demand for wood and increasing scarcity of wood led already centuries ago to the recognition, that the biological production of a forest has certain natural limits and that man can easily diminish or even endanger this wood production and, consequently, his wood supply, if he does not respect these limits and over-exploits or mis-uses the forest.

There was a gradual evolution from

careless exploitation of the forest, to an at first only
restricted exploitation, restricted for instance regarding the amount of wood which every member of a village was allowed to cut; or restricted according to age or size of the trees (diameter or girth limits);
and later to a:
careful, regular forest management on the basis of sustained yield.

This fundamental principle of sustained yield was developed in Europe centuries ago, in order to secure the wood supply of salt works or copper mines, which then had an enormous, concentrated annual wood consumption of several hundred thousand m^3 and whose wood supply had to be assured by sound forestry for 100 or 200 years in advance (this was at the same time the origin of the first and oldest forest inventories, going back to the 15th century, which have been repeated afterwards more or less regularly, in order to check the development of the growing stock and the rightness of the annual cut).

All these stages of forest utilization – from careful forest management on a sustained yield basis to 'timber mining' without care for regeneration and further growth and even to ruthless over-exploitation, exhaustion and destruction of the forest – are, as a matter of fact, still practiced on our planet.

However, there is an increasing demand for raising forest production on a sustainable basis; because forestry has to cater for a growing market it must be a main objective of forest policy, to help forestry to develop in an adequate, sound and positive way.

3.2.3 Some policy aspects

Forests can supply a wide range products and services, and a decision on priorities has to be taken. In some countries it is simply up to the forest owner to take this decision, in others it is partly or wholly a matter of forest policy and, thus, a political decision. However, forest policy will most certainly not decide for any single forest; it will fix a certain frame or set up certain general rules, which are mostly laid down in the forest law. The decision *in situ* is left – within the limits of the law – for instance to the forest owner or to the forest authority and the decision can be, that a certain forest is primarly a *commerical forest*, which should first and foremost produce wood, or a *protection forest* or a *recreation forest* etc. Some countries carry out a mapping of the forests according to their main functions; the unavoidable changes and developments are taken into account by regular ten year – or if necessary, ad hoc – revisions.

Priority decisions regarding the functions a forest has to serve, can have serious consequences, because the function, which has been given highest priority, can influence the other functions, can be a constraint to them. If for example some forests – maybe very large ones – are reserved as National Parks or Nature Reserves and any removals of wood, even of storm-felled wood, are prohibited, the biological wood production, the growth, will go on, but the technical wood production for the market or for any utilization will be nil. In other forests the technical wood production will perhaps be restricted only, because protection or recreation may have some priority but not absolute priority.

In order to ensure these 'infrastructural functions', certain constraints are fixed as a rule on the management of the forests. One of the most important ones will be a limitation in the size of clearcuts, because on clearcuts without any forest cover, the soil may be exposed to erosion; there may also be other environmental objections to such clearings. It depends, of course, on the specific conditions of a country regarding terrain, climate, especially precipitations and their seasonal distribution, endangered settlements or roads and railways, the requirements of use- and drinking water, of fresh air, of a beautiful landscape etc., what constraints on clear fellings are called for; the limits may be anything between a half or five or ten ha, but a total prohibition of clearcuts of any size does also exist, for example in Switzerland; in Austria clearcuts of more than 2 ha are forbidden and clearcuts of more than 0.5 ha need permission from the forest authority.

Constraints in the cutting age of a stand can be prescribed for two reasons: one can be the fact, that a full protective function can be expected only from older and not from very young stands, which have not got yet much stability. The second reason is very often to ensure that the forest wich is left to the following generation is at least as good and valuable, as when the present generation took it over from the former. The age-limit between immaturity and maturity can, of course, vary according to tree species, soil fertility and climate, terrain conditions etc. Even the ownership structure can be taken into account, for instance a lower age-limit might be set for farm forests. Instead of an age-limit, diameter or girth limits are in use in some countries.

Unnecessary constraints should, of course, be avoided, because they could have a negative influence on wood production.

As a matter of fact, all over the world the aim of forest policy should be, that the forests produce more wood and more services, because all that is strongly needed; of course, this aim should be attained without any over-exploitation of the existing forest; on the contrary, the state of the forests regarding productivity, stability, soundness and, if possible, even beauty should be improved.

Very much depends on the ownership structure of the forests and on the forest owner's attitudes and behaviour towards his forest, whoever the owner may be – the state, any other public agency, a company, a big or small private forest owner, a farmer etc. All these owners should know, that their forests are important for their own and the public's welfare and wellbeing, and they should all know their obligation, to treat and manage their forests well as a renewable source of goods and services. Any misuse, any neglect, any over-exploitation should be regarded as an offence against their own and the public's interest and as a breach of 'good manners'. This attitude, which is called in German 'gute Waldgesinnung', should be taught already at school (see chapter 5.1) and should be promoted by a good information service, by reports, competitions, rewards, school-forests etc.

The importance of a good extension service cannot be stressed enough! There are millions of forest owners on earth and a good many of them don't know how to treat a forest correctly; to teach them, to show them good examples and to help them, is one of the most important tasks of Forest Policy!

At the same time an efficient supervision of all forests by a competent forest authority is also necessary, an authority, which has to handle the forest law, but should act according to the slogan 'the police – your friend and helper'.

Those responsible for forest policy should do what they can to awaken the interest and goodwill of all forest owners; undue bureaucracy which stifles goodwill and initiative must be avoided; a bad taxation system may also paralyse any sound activity and development; a good taxation system on the other hand, may promote good forestry while at the same time, discouraging undesirable practices.

Regarding the supply of wood, it should not be overlooked, that the production of the forest is not the only source! One of the quickest means to increase the amount of raw material is to organize and promote the utilization of wood residues, which are generated either in the forest or by forest industries, or another possibility: the use of waste paper, which can be recycled.

On the other hand, forest policy should also try to avoid any wasteful use of wood, in order to ease the balance between supply and demand.

3.2.4 The principle of sustained yield

One of the most fundamental principles of forest policy must be that forests should be managed in a way which will maintain and where practicable, improve the productive capacity of the site and of the forest. It is also generally accepted that in all forests, whatever the ownership, management should seek to prevent too great fluctuations in the level of annual or periodic fellings and in particular, to avoid mortgaging the future by over-exploitation now. This is the idea of 'sustained yield'. It means simply to show a 'sense of responsibility for the future'. The only significant exception arises when a forest is to be converted to other land use; then it is a question of making best use of the timber by carefully phased exploitation.

Of course, by sustained yield is not meant only the amount of wood which can be harvested and used; the value of the wood must be taken into account too! It would be contrary to the principle of sustained yield, if the correct amount of wood is harvested, but all of it in the most valuable and best situated stands, near to the roads or railways. The value would then be far above average and 'bitter years' would unavoidably follow, when only minor stands and far-off ones can be felled.

However, a strictly equal level of annual cut does not fit very well with economic ups and downs. Forestry has to adapt its annual cut to market conditions and forest policy should permit the necessary 'freedom and action'. In toto however, the amount of wood removed in a, say, ten year period, should not surpass the allowable cut for that period. For farm forests a longer period of, say, 30 years may be appropriate, that is the average time for a farmer to be 'the boss'; this should be a guiding line, not to take more than is any generation's right share.

The problem is perhaps most difficult with regard to 'collective forests' (Gemeinschaftswald), because there are many owners, who share not only the property of the forest, but also the annual cut. However, in order to adapt the removals as far as possible to market conditions, all part-owners, who do not need their share in a year when prices are poor should be allowed to postpone their share to one of the following years. Of course, this means some administrative work, a kind of book-keeping for every member. An inverse procedure can be applied in years of good wood prices insofar that interested members can take not only their left-over share, but already next year's share too or even more as an anticipation for coming years.

The principle of sustained yield and of an allowable or potential cut should, of course, be applied to a country's forestry as whole also, though this will, as a rule, not be more than an orientation and corrective for policy measures aimed at ensuring a satisfactory regular supply of wood for industry and domestic use and reasonably secure and continuous employment for those engaged in forest work.

The application of the sustained yield principle presents special problems when management is first introduced into virgin forests or other forests with a large proportion of mature or over mature timber. To manage these forests on the sustained yield principle means to cut the old growth at a rate which permits the young growth which replaces it, (whether by planting or by natural regeneration) to develop so that no serious gap in yield will occur when the old growth is exhausted. Usually the second growth trees develop faster and can be harvested at a much younger age than the original trees. This of course applies particularly when new species are introduced which grow more rapidly than those of the original forest. Such a species change usually brings about also major changes in the type of forest product, for example, high quality tropical hardwoods may be replaced by utility saw logs and pulpwood from plantations of pine. Under these conditions there may be a sustained, or even a sustainable increase in money yield but not of the original product. It is clearly an issue of some importance to decide the period over which it would be desirable to utilize existing old growth and to take appropriate measures to secure that objective. This issue may arise both at national policy level and at local management level. The sustained yield principle can of course be applied without difficulty to the new crops.

A different problem arises in the opposite situation when instead of starting with an old forest, a country or a district or even a single owner begins to afforest bare land; large scale examples of this are found in many parts of the world. Here the yield gradually builds up in quantity and changes in quality from the time when the first plantations reach the first thinning stage until there is a complete series of age classes when sustained yield really begins. In a situation of rising yields, forecasts are particularly important so that new industries may be established in step with the rising supplies.

There are, however, also cases where the principle of sustained yield does not apply. If for instance, the existing forests are in a bad state, maybe too old, without any growth or poorly stocked or damaged by game or wind and snow or consisting of unsuited species, then it would not make sense to manage these forests as if they were normal valuable ones. These forests should be converted as fast as possible into better ones but avoiding too big clearcuts and taking any other necessary environmental precautions.

3.2.5 The rotation period and its function

The rotation period is the average cutting age of the trees or stands in a given forest unit (estate, district, country etc.), however in some cases it is not the age which is decisive; it can be the size of the trees or their growth. The length of the rotation period is inter alia a guideline for the dimensions the trees can attain in this span of time, and these dimensions are on the other hand often decisive, for the uses to which the wood can be put.

To fix the rotation period is normally a decision of the forest owner, but it can also be regulated by law; in either case it is up to the forest authority to show the possibilities and the advantages and disadvantages and to provide the decision aids. For most forests in Central Europe the rotation period will be between 70 and 120 years; for valuable, high quality hardwoods it can be as long as 200 years and more, as for example, for the very high-priced oaks in France or Western Germany. Plantations of fast growing species in many parts of the world (see the following section) are often managed with a rotation period of 30 years or less and the modern, so-called 'energy plantations' will be run on a rotation of three years or even less.

Even in the tropics, large dimensioned wood of high quality generally needs a long rotation period, which can be afforded only by public forest owners or big private enterprises because of the very high capital, which is tied up in the growing stock of these forests. Wood of such kind needs on the one side a very long time to grow, but has on the other side always a good and stable market, usually at high prices; a single tree may fetch several thousand dollars. Large dimension alone is, of course, not sufficient; it has to be connected with high quality. As a matter of fact, trees that are too large, if they are not of first quality, may fetch a lower price, because they are difficult and costly to handle and often too thick for the normal frame saws, which are widely in use in some parts of the world. There exists often a 'turning point' at which the price-size gradient which normally mounts with the dimension, begins to fall.

'Mass-produced wood' needs, as a rule, a much shorter rotation period and, thus, much less growing stock and less invested capital, but the market for this wood has sharp ups and downs and this can bring the producers in serious difficulties, the more so, if the buyer is a strong monopolist and forestry is the weaker part. However, if for example, a pulpmill owns forest land, it might be quite logical, that the production goal will be small dimensioned pulpwood, which can be produced in a rather short rotation period. For woodland owners other than forest industries, it is usually inadvisable to put 'all eggs into one basket', especially a pulpwood basket. By growing trees to sawlog size there is a much wider choice of markets because saw logs can be pulped, but logs of too small dimensions cannot be sawn. Moreover, a given volume of timber made up of medium sized logs costs far less to harvest and extract because there are fewer logs to handle.

3.2.6 Natural forests and plantations

Real virgin forests, as nature has built them without any influence of man, are getting fewer and fewer. It has already become urgent to protect some of them by creating National Parks or Nature Reserves, in order to reserve some for future generations. On millions of hectares the former virgin forests have been replaced by 'man-made forests'; there are, however, big differences:

Man-made forests can be and often are, 'near to nature'; they are somehow related to the former natural virgin forests, but they have as a rule an 'improved' species composition, improved insofar as species, which man wants most, have been given more (sometimes too much.) space, but as a whole they look somewhat like a natural forest and they have a good many of the valuable properties of a natural forest, such as stability, soundness, high productivity etc. This kind of man-made forest is widespread in the boreal region.

Other kinds of man-made forests are less 'near to nature', being for instance, monocultures of one single species – and they are in the main a result of a clearcut followed by planting. Forests of this kind tend to be more prone to suffer from pests and wind; these dangers may be reduced bu regular thinnings which strengthen the most promising trees and thus improve the stability of the stand and its resistance to wind and disease; at the same time the growth will be accelerated. Even-aged plantations of a single species are very different from what is generally understood by the term 'forest'. They bear in fact more resemblance to agricultural crops, to fields of maize or sugar cane; monocultures, artificially created after a thorough soil working, may be several times fertilized or treated with pesticides and finally harvested after a short rotation period. Under certain circumstances, these wood plantations are the best, perhaps the only way to produce big amounts of low-priced utility wood, for industry as well as for domestic use (firewood, building poles etc.), but they generally contribute little, if anything, to the prevention of erosion. In extreme cases they may degrade the soil, aggravate erosion and spoil the landscape.

The nowadays widely discussed 'energy forests or energy plantations' are based on the same principle as the former coppice forests, which were wide-spread in the 'old-world'; they were managed to produce mainly fuelwood in a short rotation period of 5 or 10 or 15 years. These forests were regenerated naturally after clearfelling with very sharp axes, by sprouts from the stumps. The 'energy forests', will be managed in a much shorter rotation period of one or two or three years only; their regeneration will also be a natural one by the sprouts from the stumps. If properly managed, they can help indeed to increase the energy supply, mainly however at the local level only in rural areas. The extration of nutrients by the frequently repeated harvest of most of the biomass (wood, bark, twigs) is, however, very high, so that, except possibly on very good sites, fertilizing may become necessary, in order to make up for the losses of nutrients; in this case the final energy balance might become at least questionable. An alternative to energy plantations which is favoured my many foresters is to make more use of branches and bark in conventional forests and plantations, of residues from wood processing (sawdust, bark, slabs, etc.), and of recycled wood (from demolished buildings etc.).

The issue between 'forests near nature' and even-aged plantations of a single species managed more like agricultural crops boils down to this: on a purely financial view and to meet immediate market demands, plantations may be best and it may even pay to reduce or omit tending operations such as thinnings; but these culculations take no account of the possible environmental disadvantages wich are difficult or even impossible to quantify in money terms (see chapter 4). The policy decision with which the forest owner, public or private, is faced, is how much revenue he or she is prepared to forego or what extra costs he or she is prepared to incur in order to meet environmental desiderata. Governments are not always very logical in approaching this problem. On the one hand they may almost force owners public and private to take a very narrow financial view (e.g. by inadequate budgets for state forests, ill considered taxation in private forests); on the other hand, they may spend considerable sums on nature conservation in their environmentel budgets (tree planting schemes, nature parks etc.). It might sometimes be more cost effective to spend a little less on some purely environmental programmes and to do more to promote sensible forest management.

3.2.7 Permanent or interrupted biological production

If a stand is harvested and all trees are removed by a clearcut, growth goes down to zero and will resume only after a number of years, when the area is again stocked with young trees, either planted or seeded naturally. These years of no or very little growth are, of course, a loss in biological production and they may even be a loss to the site, because valuable nutrients can be washed out and the precious nitrogen can be used up by the often very luxuriant vegetation on sites that have been clear cut.

Already centuries ago, forest owning farmers in Switzerland, Southern Germany and parts of Autria strictly avoided therefore any clearcutting; they managed their forests by a 'selective cutting system' and felled single trees only or small groups of trees, which were already mature and had a sound undergrowth of young trees: the next forestry generation! Scientific forestry elaborated then various different, but in principle similar systems called shelterwood systems. (Plenterwald, Gruppenplenterung, Schirmschlagbetrieb, Femelschlagbetrieb etc.), all of them based on natural regeneration under the old trees. The latter have, of course, to be felled very carefully, in order not to damage or to destroy the young trees. In a sense these systems bring about two forests at the same place and at the same time – the old one and the young regeneration! Such an uninterrupted biological production is, of course, very advantageous, but to carry it out sounds perhaps easier than it is in practice. The various kinds of selective cutting are more difficult and usually also more expensive than the easy clearcutting system and that is why most of the world's forests are managed by the clearcutting system; the management can often not be as intensive, as it has to be with a selective cutting system, because of lack of the necessary infrastructure (no adequate road network) or lack of personnel (foresters and forest workers) or lack of knowledge and skill; the regeneration of stands consisting of light-demanding species like poplar and willow, pine or larch, under older trees is more difficult but possible under favourable conditions. It should, however, never be overlooked, that a clearcut removes the forest cover totally for a number of years and that soil erosion by water or wind may occur or that even the water regime as a whole can be disturbed, because there are no trees to absorb water from the soil and transpire it to the atmosphere.

3.2.8 How to increase wood production

The 'hunger for wood' is worldwide and steadily increasing and there are doubts how this hunger can be stilled especially after 2000.

Those responsible for forest policy have therefore the very important, but by no means easy task, to examine and publicize the posibilities how to increase wood production and how to get things really going; dozens of excellent concepts and plans have NOT been carried out! Most tasks in forest policy as of other policies are connnected with persuading and convincing the people, who are responsible for the forest – the political leaders, the organizations, the forest owners and foresters including the forest workers and last but not least, the broad public. For some of the necessary or at least recommendable measures it will not always be easy to get public support, because naturalists, environmentalists and con-servationists often object against measures to be applied in forestry, which are applied on a 10,000 times greater scale in agriculture (fertilizers) or in households and agriculture (insecticides); in forestry they think all this should not be done, forests should be and remain an oasis of undisturbed naturalism. They seem to forget that forestry is an important branch of the economy and that uncountable jobs are connected with forestry and forest industries.

Sometimes it is necessary to take a difficult problem not at its core, where nothing but obstacles can be expected, but to take it from the easiest side and to encircle the difficult core, until this too will be 'ripe'.

There are four main ways of increasing the availability of wood:

– to harvest more wood;
– to make better use of the wood that is harvested;
– to raise the rate of growth in existing forests;
– to afforest bare land.

It is a common misconception that raising wood production is necessarily only possible in the long term. The first two of the above methods can give immediate results.

Harvesting more wood. Even in countries with a long history of forest manage-ment there are forests not in use, because they are as yet inaccessible. They can be opened up for forest management by building forest roads. This should be done on the basis of a complete road plan and not piece meal to gain access to particular stands. Although road building, especially in mountainous terrain, is costly, there is an early income from the timber to which the road gives access and some of these roads may also serve the general purpose of improving communications in remote rural areas. Also in forests already in use it may be possible or even desirable to harvest more wood. An often neglected task is to carry out regular and sufficiently heavy thinnings; these thinnings may not always be profitable in themselves but they get more wood to the market, they enable the trees that are left to grow faster and under certain conditions to become more resistant to damage by wind, snow and disease; thinned stands are also usually more pleasing to the eye.

Better use of the wood that is harvested can be made in various ways: more careful grading which enables each piece of wood to be put to its best possible use, more efficient processing which reduces waste, preservative treatment of timber to prevent decay, and increase resistance to fire, the use of improved wood products. To quote a few examples: modern sawmills not only reduce waste by sawing more accurately and getting the maximum yield in quantity and quality from each log, but they also produce residues which can be used for pulping or the manufacture of particle board. Preservative treatments can more than double the life of timber in contact with the ground. Recycled waste paper is already an important component of many pulp products, but the scope for more recycling is great; the main limiting factor is the organization and cost of collection.

The rate of growth in existing forests and the quality of the timber produced *can be raised* by more intensive silviculture (favouring of certain species, thinnings etc.), and in the longer term by tree breeding. The application of fertilizers and drainage must also be mentioned as possibilities. Unfortunately, in many parts of the world the growth potential is reduced because the best trees are always exploited and the inferior ones left to grow on. In extreme cases the forest is eventually destroyed. Avoidable production losses also are incurred through uncontrolled grazing, careless fires, damage by excessive populations of game (deer, elephants etc.). There is of course also the loss through the conversion of forest land to agriculture or to urban use.

The afforestation of bare land has already added to the forest areas of many countries but even in regions with a large population there is still much land that is more suitable for forestry than for agriculture and not needed for agriculture. According to estimates by the EEC Commission (1979), there are some 4 to 5 million ha in the EEC alone.

The detailed decisions on how best to increase production and make better use of wood must rest with the forest owners and industries immediately concerned, but it is up to governments and their forest authorities to make known what can and should be done, to remove obstacles as far as may be practicable and to provide incentives where appropriate. These are the policy issues that need attention.

3.3 Marketing

3.3.1 The market economies

General considerations
The most essential feature of a market economy is the fact, that the market, which is characterized by offer, demand and price, is regulated automatically by the

so-called market forces without any direct influence of the state, of the government or of any authority. However, it would be wrong to assume, that market economy means a 100% free system. There are at any rate the framework conditions, which are fixed by the state or given by nature or by outside factors like policies of other states. These framework conditions influence for instance the connection to the world market or to neighbouring countries, but influence as well the position of wood and wood products in the country itself.

The competition with other countries can for instance be mitigated by customs or even excluded by import or export prohibitions; some countries prohibit, for example, any exports of roundwood or impede them by export taxes, others try to protect their industry by prohibiting imports, mainly of finished goods like furniture or doors and windows or by hindering them by customs. Currency agreements can also be an efficient means to influence foreign trade; the same is valid for credits which a country gives to others, in order to improve and augment its exports, i.e. to enable these countries to buy the goods.

Within the borders of a country, in its own economy, free competition is a ruling principle of market economy, but policy in general determines also here the framework conditions, mostly by credit measures, interest rates, promotion of building activities etc. To a certain degree the state can influence directly the market, for instance, by the removals and offers of roundwood from the state forests or by its own economic activities such as construction of roads and motorways or of public buildings like schools, hospitals etc. and by preferring the use of wood and wood products or by limiting it. However, free competition and free price formation are normally warranted; there may be some exceptions for goods of vital importance as for instance food, whose prices are often regulated by the government, or regarding monopolies and cartels, which could otherwise become too strong a 'market power'.

Because of this principle of free competition it is as a rule not permitted to direct the roundwood supply with preference to local industries, which could be a subvention of these industries by forestry; this is not permitted within the country itself, nor, as a rule within the various 'common markets' like EEC, EFTA etc. Good relations between the enterprises at the local level make it, however, feasible to prefer a certain partner, as is pointed out in the subparagraph on auctions (see page 97).

The free market economy in its extreme form was typical of the times of high capitalism. The political and socio-economical development led at first in the Federal Republic of Germany soon after World War II to the modern so-called 'social market economy', but now there are signs that a further step will come towards the 'human market economy'. The difference between social and human market economy is in the main, that the social market economy tries to avoid or to mitigate hardship for weaker social classes or social interests, whereas the human market economy takes into consideration also possible harmful environmentel and ecological effects which should be avoided or minimized. In both cases the framework conditions are set by parliament and government, but a good many details can be agreed upon by the market partners themselves or by their organizations; they are all interested in a sound and undisturbed market and the state should according to the principle of subsidiarity only interfere when the direct partners are unable to carry out the necessary measures. Forest policy should therefore seek to influence the forestry and forest industry interests concerned to discuss their problems, so that they themselves may find the best solution and come to a sensible and fair agreement.

One of the most fundamental instruments of any kind of market economy is the so-called price-mechanism. This mechanism steers the behaviour of suppliers, the forest owners, regarding offer and of buyers, forest industry or wood consumers, with regard to demand. The price, which develops and changes automatically according to the relation of offer and demand, has a decisive informative function for the participants of the market. If for instance, the relation between offer and demand is not well balanced, demand being bigger than offer, prices will increase automatically; this price increase will act as an incentive for suppliers to increase their offer of supplies; this will work, until a balance between offer and demand is again obtained. The contrary will happen, when supply is bigger than demand; prices will drop and this will lead to a market-conform behaviour of the suppliers by diminishing the volume of goods they offer. The reaction of forestry will therefore be to adapt the offer, ie fellings and removals, to the level of demand, whereas the reaction of industry is mainly dictated by the market of their products; but industry can react also by increasing or decreasing the roundwood stocks.

Very important and valuable are, of course, good contacts between forestry and industry at the local, national and international level, in order to discuss the situation and likely development of the market; at the ECE-level the annual meetings of the Timber Committee are an excellent example; in Austria or Switzerland similar discussions are held on the national level.

The self-regulating forces of the market work as a rule very well; offer and demand correspond normally satisfactorily and no heavy disturbances of the market will occur, unless . . .

Unless for instance, 'force majeure' comes into the play: if, for example, the supply is augmented by heavy windblow or snowbreak. This wood has to be worked up and brought to the market, in order to avoid a dangerous epidemic of bark beetle or a deterioration of the wood. Then offer and demand no longer are in balance, prices will drop sharply and the economic loss to forestry can be a tremendous one. This disturbance of the market can even lead to bigger than normal removals and sales of wood by overcuttings, which are from the point of view of forestry policy absolutely undesirable; forest owners, because of the lower price of wood may be forced to cut and sell more than normal in order to secure the necessary minimum income and this is, of course, not only a disadvantage to their own forest, but also to the market because the offer, which is already too big, is further increased. Even 'panic-sales' by forest owners can occur, if they are afraid that wood prices could continue to drop.

In all these cases it is a very important task of those responsible for forest policy to act according to a kind of 'mobilization plan' (see 5.1), prepared and agreed upon beforehand by forest and industry. A quick and efficient information service is then vital, so that both sides know, what to do. Forest owners who are not affected by the calamity should be persuaded to reduce immediately their removals and sales, in order not to increase the offer, which is already too big. Forest owners and industry have to be helped to increase their roundwood stocks by technical and financial aids or to get rid of the surplus for instance by roundwood exports, which may normally be considered undesirable. Similar contingency planning is desirable as a precaution against an unexpected sudden drop in demand which could be caused by, for example, the closure of major pulpmills as happened in the UK.

Methods of marketing

In market economies the formation of prices is governed by the relation of offer and demand. There are, however, different possibilities, how offer and demand come to meet.

In a market economy, many sales are by 'negotiation' i.e.seller and buyer meet and agree after some bargaining, on the price and all other details. This works as a rule quite well, as long as the two partners are of similar strength. It could, however, go wrong, if one of the partners is much stronger than the other and can, thus, influence the price to his advantage; for example, a small forest owner with almost no experience of the market and the representative of a big forest industry will not be equal partners unless, as is the case in many countries, the organizations representing each side arrange a bilateral price agreement at their much higher association-level, where the weakness of the small owner is compensated by the weight of his organization. This agreement can, for instance, provide a price frame or price-range for sales and purchases according to quantity and quality. The aim is to protect the weaker against the stronger and to give him more or less the equal strength through his organization.

Big forest owners may not need this support; they can even think of auctioning their wood so as to sell it to the best, highest bidder. Sales from state forests in some countries are made in this way. The auction may work 'from the bottom up', that is beginning with a modest price and letting the price increase according to the price-offers of the potential buyers, or 'from the peak down', that is the offer begins with an extremely high price and goes down, until a bid is given; either way, the highest bidder gets the purchase. An advantage of the system is that it leads to 'market transparency'. Everybody can find out the prices paid at an auction. In some countries, e.g. France and the UK, auction results are published in the forestry journals. On the other hand, a disadvantage of this system can be that the best bidder, who gets the wood, will as a rule be every time somebody else, that is the buyer, who needs the wood this time most urgently and is prepared or even forced, to bid the highest price. Something like a friendly, lasting partnership between seller and buyer will, thus, not come into existence. Such a friendly, lasting partnership may, however, be not only very useful, but sometimes even life-saving, especially in case of heavy market disturbances. A good partner will not let the other down; next time he will perhaps be the one, who needs the other's help.

Auctions can have one more disadvantage; if the interested buyers are assembled and personally present at the auction, prestige, grudge and jealousy can play a rather important role and the price, which is being built under these circumstances, can be an unsound, exaggerated, over-heated one, which disturbs the whole market and leads perhaps to political consequences, not to speak of unkind comments in the press.

This can, however, be avoided, if the auction is carried out 'in writing' as 'sale by sealed tender', that is written price offers in closed letters to the seller, which are opened at a certain date and the best bidder found. The disadvantage of no permanent partnership coming into existence is here, of course, the same. However, if the market is very uncertain, it has the advantage, that the prices offered need not to be disclosed, of if there are only few interested buyers, the formation of price-rings is more difficult than at auctions.

In some countries sales of wood on the stump without any subsequent measurement of the felled trees are prohibited. This is mainly as a precaution against overcutting and to protect small woodland owners with little knowledge of timber measurement against unscrupulous timber merchants. Other countries have no such restrictions. Very much depends of course on the ownership structure of forests in a country and on local traditions.

Regarding the measurement of wood, the aim should be to avoid unnecessary labour and to simplify the work burden to both partners. For instance, to change from measuring individual logs to calculating the volume by weighing the whole load of wood on a lorry or wagon at once, is of course an advantage for both sides, if it is carried out correctly and the degree of humidity is well taken into account. Weighing is now widely used for pulpwood and to a lesser extent for utility sawlogs. It is obviously less suitable for high grade logs, where the quality and unit price may vary considerably from log to log. Somewhat similar considerations apply to the question of measurement 'in bark' or barked. To remove the bark by hand, as it was done up to one or two decades ago, is not only a very strenuous labour, but also very costly. Now it is done almost everywhere by machinery at the sawmill or pulp mill and the wood is, therefore, sold and bought in bark at the forest; here also, a partnership agreement is necessary on how to estimate and allow for the volume of bark and it is up to the forest authority to see to it that the partners come to such an agreement.

As wood is a variable raw material, the grading of sawlogs and sawnwood constitutes a very important aspect of marketing; it enables the vendor to specify more precisely what he is selling and the perchaser to know what he is buying. Until recently, grading was mainly based on visual characters (size and frequency of knots, width of rings etc.); for sawnwood these subjective methods are now being replaced by more objective methods of mechanical stress grading which are more closely related to likely performance in use.

Some forest owners prefer to sell their trees on the stump (ie. standing) while others prefer to sell the trees after they have been felled and extracted to road or rail; some may even arrange for the transport and sell at 'factory gate'. For small farmers and other small woodland owners, earning additional money by harvesting trees may be an important element in total income; moreover, forest work can often be done at times of the year when there is little farm work. The risk of damage to the trees that are left standing is also usually less if the felling and extraction is done by the owner or at least by workers directly employed by him. That is why some state forest administrations employ their own work force.

The offer of long term supply contracts have sometimes been a useful incentive for the establishment of major new forest industries. An initial price is negotiated which is then varied according to some pre-arranged price index or combination of indices, e.g. wholesale price index, price index of the products of the particular industry, wage level index. Even with a carefully chosen system of annual automatic price adjustments unforeseen circumstances may arise which unduly favour either the seller or the buyer. Some provision for re-negotiation and premature termination of a contract, if re-negotiation leads to no agreement, is therefore desirable. If too large a proportion of all supplies is committed by such contracts a disortion of the market for the remaining produce offered is almost bound to accur.

Sales where the seller simply fixes the price are common only for retail sales of minor forest products such as fence posts or Christmas trees.

A common way of opening up large tracts of forest for management and forest industrial development has been to offer concessions. The forest owner, usually the state, enters into a long term contract with a forest industry whereby the industry manages a forest area and has exclusive rights to all the produce from it in return for a royalty, usually a fixed sum for every unit of volume of timber removed from the forest. The advantage to the forest owner is that the forest is put into use and that he obtains an income without having to incur initial capital expenditure on roads etc. The industry has the benefit of a secure supply of raw material under its own control. Some such arrangements have worked very well, but unless the forest owner exercises great care in drawing up the initial agreement and in monitoring the subsequent performance by the industry, the forest may be destroyed and the revenue to the owner very small. It has happened all too often not only in developing countries, but also in developed countries that the forest authority or other government department concerned is too inexperienced in commercial matters to negotiate reasonable terms for a concession with a major industry. Among the points to watch are the indexing of royalties, the prevention of cutting only the best trees and insistence on sufficient capital investment by the industry in long term infrastructure (roads etc.) and industrial plant to give it a vested interest in a sustained yield from the forest.

As far as external trade is concerned, every country will try to export finished or semi-finished goods and to import, if necessary and possible, raw material, ie roundwood or chips in order to maximize export earnings, minimize the costs of imports and to give more employment to its own people. The problem is especially important for developing countries with surplus labour and limited export possibilities. Experience suggests that there are dangers for these countries in trying to develop sophisticated forest industries too fast, but they should at least make a start. Helping developing countries along this road of industrial development is one of the most effective forms of aid which developed countries can give.

3.3.2 The centrally planned economies

General considerations
In centrally planned economies, the greater part of the means of production is social property, owned by the state and the co-operatives. This circumstance makes the comprehensive planning of the entire process of economic and social development possible and necessary at the same time.

The objectives of the fundamental political, social and economic development are determined by the ruling parties, governments and parliaments. One of the most important tools for the realization of the objectives is planning, which encompasses all substantial factors of economic growth, the natural resources, the means of production and the labour force.

Directives for planning are issued by the supreme party and state organs. These directives usually contain the objectives relating to the general development of society, economic, cultural and social needs, as well as international relations. But the concrete plans are worked out by the central planning organs.

In the course of development, for the Soviet Union from the October Revolution, for the countries of Eastern Europe after the end of World War II, up to the mid-1960's, the plans in these countries were usually characterized by the fact that most targets were of compulsory or obligatory nature, whether in the form of direct instructions and directives, or in indirect ways, down from the government level through the ministries to the production unit. The obligatory planned figures stated the scope within which a given economic unit could carry out its activities.

In centrally planned economies the principal method of ensuring consistency in planning is the employment of various balances between the targets of the plan and the available resources. These balances present on the national economic level, the needs and resources in respect of material goods, values and labour.

The system of national economic balances embraces the entire developmental process, follows the movement of products from production to consumption, the movement of labour and the increase of national wealth.

The national economic balance usually consists of the following three principal parts:

- social product balance;
- national income balance;
- integrated labour balances of the national economy.

In practice in the planned economies three types of plans are generally used: the long-term plans (15-20 year), the medium-term plans (5 year) and the annual plans. National economic balances are set up for each type of plans.

The balances of the wood products sector fall within the framework of the social product balance. In the 1950's almost everywhere in the centrally planned economies the forest product balances contained a long, very detailed list of products, including the forest products (veneer logs and sawlogs, pitprops, pulpwood, poles, fuelwood etc. by species, dimension, quality groups) and the processed wood products (veneer, sawnwood the various wood-based panels, pulp, paper, furniture, etc.), also in different groupings.

The different forest product balances have contained the sources/production, import/and the distribution to the different consumers/factories, export, individual consumers. In this situation the commercial enterprises fulfilled only the function of distribution of the products, in the field of forest products trade also.

Up to the second half of the sixties the word 'marketing' did not have any real meaning in the centrally planned economies. For better understanding the situation it seems to be useful to cast a very short historical glance backward.

After World War II, when the Eastern European countries joined the centrally planned economy system, led by the Soviet Union, there existed in these countries a ration card system for many consumer goods. Up to the mid nineteen fifties, in the Soviet Union and in the other centrally planned economies a 'market' simply did not exist, which has regulated the production and consumption in the sense as it is used in the market economy terminology.

When these countries in the fifties began to turn to the peaceful economic development, it was necessary to lay down the theoretical basis of building up the socialist economy and society. In the text book 'Political Economics' published in the Soviet Union in 1954 we find among the important theoretical aspects of managing a socialist economy the following, which are relevant to our work:

- the direction and functioning of the economy is based on the state plan, which contains compulsory production distribution figures for all industrial, agricultural and commercial enterprises;
- in the socialist society, trade is limited mainly to the group of personal con-

sumer goods, which are produced in a quantity determined by the state plan;
- the land resources are in state ownership, they cannot be purchased, sold, or hired;
- the state owned enterprises, factories, mines, electric power stations etc., cannot be a subject of a sale contract;
- the means of production, engines, tools, coal, oil etc., produced in the state owned sector are distributed among the state enterprises according to the annual state plan;
- foreign trade is a state monopoly;
- the state farms, forests, co-operatives are obliged to sell their products at fixed prices to the relevant state commercial enterprises.

On this theoretical basis the market did not have any regulating or influencing role in managing the economic life. The economic life is regulated by the plan; the relevant bodies, concerned with planning activities in the centrally planned economies are the Planning Committees; they are subordinate to the Council of Ministers, and replace generally the role of the market.

Under the word 'market' at that time was generally understood the so-called 'free market', a place in the villages and towns, where the members of the agricultural co-operatives sold at free market prices their products produced on their limited household plots. Members of a co-operative are entitled to have a household plot of about $\frac{1}{2}$ to 1 hectare to cover their own needs in food.

During the fifties and sixties the economy of the centrally planned economies was generally managed on the basis mentioned above.

In the 1960's, the centrally planned economies approached or attained the phase of transition from the extensive phase of economic development to the intensive phase; industrial production doubled or even trebled in a decade.

The methods used with success during the war economy and the extensive phase of development had been becoming more and more inefficient. The enormous number of compulsory production figures led to an increasing bureaucratization and diminished to some extent the responsibility and initiative of the managers of the factories. From this time the different planned economies developed and improved their planning and managing systems.

Every country developed its system in his own way, taking into consideration the specific conditions of the country. The systems which were nearly uniform in the centrally planned economies in the fifties have been becoming rather different from the end of the sixties. Some countries have taken only cautious steps, others have made substantial changes. The direction in which generally all countries are developing their system is to diminish the number of compulsory production figures, to increase the responsibility and initiative of the managers of the factories and thus to raise the efficiency of the whole economy and to promote the independence of the factories.

This is not the place to characterize the now existing economic regulating system of all the different centrally planned economies; instead, we take only one example. In Hungary, a general reform was introduced on 1 January 1968. Among the other countries only Poland has carried out a general reform; the others continue to use obligarory planning targets as the main tool for regulating the economy. Poland's new system has some features in common with that of Hungary but it is too early to judge results.

The main features of the so-called Hungarian 'New Economic Mechanism', the methods of which aroused great attention throughout the world, can be characterized as follows:

- the obligatory figures have been abolished. The five year plans contain general indicators of development of society, economic, cultural and social life. The five year plans are endorsed by the Parliament and are obligatory for the government. The government directs the whole economy, mainly through indirect economic incentives.
- the enterprises have become independent and have got on the one hand the right to determine their production targets and on the other hand the responsibility to manage the enterprises raising continuously their efficiency and providing maximum profit.
- among the economic incentives, which motivate the activity of the enterprises, one of the most important is the price system. The system, which was introduced in January 1968 implements a so-called mixed price system with three categories of prices. There are *fixed prices*, mainly in the case of the most important raw materials and consumer goods, *limited prices* and *unrestricted prices*, mainly for industrial finished products, for vegetables, fruits etc. Other incentives, such as taxes, import duties, subsidies in limited and transitional cases, etc. are also important elements of the regulating system.
- the new economic mechanism recognizes the important role of the market, which has to give the necessary information for the managers of enterprises in decision-making. This market is not an absolutely free market (nowhere is it that nowadays), it is a market influenced by the main indirect incentives of the government (prices, taxes, import duties, specific incentives etc.).

The new economic mechanism has been working effectively for more than a decade. The world economic recession and internal effects require the continuous improvement of the system. For example, the improvement of the balance of the economy significantly depends on whether the changes in foreign trade conditions are foreseen and in due time. All that necessitates a considerable improvement in the market research activities of the foreign trading agencies and in forecasting work.

Another experience is that the increasing role of the human factors demands that the social and public organizations be given a greater role in planning and necessitates, therefore, improved relations between the national economic planning organs and the social and public organizations, such as the trade unions.

The other centrally planned economies have not gone as far in changing their systems as Hungary, but all of them are continuously improving their systems according to their historical development and specific conditions. This is the general framework in which the marketing of forest products finds its place.

Methods of marketing and of influencing supply and demand

From the previous paragraphs it is evident that the economic mechanism of the different centrally planned economies differs from country to country. Much greater are the differences in the absolute and relative magnitude of the forest area. First of all the Soviet Union differs significantly from the other countries, because it has one of the greatest forest reserves in the world (3.5 ha forest and other wooded land per caput) and can therefore export a considerable volume of timber products in addition to meeting its own requirements.

In two countries, Hungary and the German Democratic Republic, the forest area is smaller than 0.2 ha per caput which means that a considerable part of their timber requirements has to be met by imports.

The other countries, Bulgaria, Czechoslovakia, Yugoslavia, Poland and Romania, with between 0.2 and 0.5 ha of forest area per caput are on the edge of self-sufficiency, with some exports, but it is likely that further increases in demand will reduce exports.

This explains why the marketing systems differ from country to country in the centrally planned economies.

Forest policy's influence on *supply* is unambiguous; in the centrally planned economies as elsewhere in the world the common attitude is that:

– the basic objective of forest policy in these countries is that of increasing the production of forests and forest products;
– logged areas must be reforested everywhere, the forest area must be increased in the forest-poor countries;
– forests must be developed towards more intensive management in order that increment should increase regularly both at national and local levels;
– the species composition of forests must be improved, first of all by the plantation of fast-growing species, such as poplar and conifers;
– the multiple role of forests should be developed for the benefit of society.

The *means* of realizing the aims of forest policy are the following on the supply side of marketing of forest products:

– The great majority of the forests are in state or co-operative ownership, and this fact gives an effective opportunity to the authorities to implement the aims of the forest policy through direct regulation;
– in centrally planned economies, the drawing up of forest inventories is con-

sidered to be of fundamental importance, and measures are taken for determining the forest area, the forest resources, the wood species composition, the age group distribution and other necessary data. This work has been completed practically everywhere and these countries now have sufficient data on their forest stands;
- Forest inventories serve as the basis for drawing up management plans. Such plans are being prepared for regulating the work in the forest for longer periods. Management plans are usually drawn up for periods of 10 years, but data are supplied also for timber volume to be exploited in the next 20 years. The management plans contain the allowable cut for a decade on the basis of sustained yield. The management plans provide the necessary basis for forest policies and planning. In the centrally planned economies the drawing up of management plans is obligatory;
- on the basis of the public ownership on the one hand and of the data of the management plans on the other hand, the different plans, first of all the five-year plans, contain the concrete measures necessary to fulfil the aims of forest policy to raise the supply in quantitative and qualitative terms.

Forest policy's influence *on demand* varies from country to country in the centrally planned economies. Let us take first the *mechanism* of influencing the demand for forest products. In most countries the extent and composition of timber consumption are determined from detailed balance calculations for each of the most important user branches.

Requirements in the user branches are determined from aggregate norms, e.g. the consumption of sawn softwood for building activity in terms of 1 million units of national currency, or the consumption of pitprops per 1000 tons of coal mined.

The volumes of consumption of forest products are limited and directly connected with the planned targets of the user branches. These branches are usually represented by the different ministries which distribute the fixed amount of forest products among the factories according to their production plans. This system is based on the obligatory planned figures and the disposable amount of forest products to be distributed. In this case the indirect economic regulators have a minor importance and the aims of forest policy are carried out mainly by obligatory means.

In some countries, especially in Hungary, where the system of obligatory figures was abolished in 1968, the influence of demand on forest products is achieved mainly through the market, which in turn is influenced by the different economic regulators. In some countries, a mixed system exists in which some important products are distributed, while some others reach the consumer through the market.

The *targets* of forest policy generally influence the demand in the direction of inhibiting the increase of consumption of forest products. This is the general feature both in the importing and in the exporting countries. The main reason is that these countries seek to increase their forest resources, while the allowable cut limits the level of annual fellings. Taking into consideration that the importing countries try to keep their imports as low as possible to save hard currency and the exporting countries to maintain their exports as high as possible to earn hard currency, in both categories of countries forest products are commodities in demand.

This is the reason why substitution of forest products is encouraged by materials which are more abundant, such as concrete, steel etc.

Recently, the world economic recession, which is to some extent influencing the economic development of the centrally planned economies also, is causing some changes in this respect. We can recognize some relaxation in the timber demand and supply in these countries because of the problems of export to the market economies and a decrease in the rate of economic growth in the planned economies, whilst the production of forest products is increasing further.

The more 'modern' wood products are now appearing in such fields as buildings for animal husbandry, week-end houses etc., where wood nearly disappeared over the past quarter century.

3.4 Forest policy and forest industry

3.4.1 Introduction

Forestry and forest industry are, as a rule, closely connected. Forestry is the producer of the raw material wood and industry is the processor and converter of this raw material into a big number of products.

Imports and exports of wood in the rough, can, of course, loosen this interdependence, even to a point where either most of the wood needed by industry is being imported (for example: Japan) of where roundwood is exported and transformed elsewhere (for instance: some developing countries).

Normally, any country, whether developed or developing, will try to make the best use of its forest resource and will, therefore, want to have an adequate forest industry potential, which is sufficient to process most of the indigenous, and possibly also imported roundwood. The development of appropriate forest industries is thus a key concern of forest policy as well as of industrial policy and both foresters and forest industrialists have a common interest in the best possible forecasts of supply and consumption of wood and wood products as far ahead as may be practicable.

3.4.2 Forecasting supply and consumption

There exist a host of simple, complicated or extremely sophisticated methods of forecasting. Which ones can be used, depend very much on the data base a country has at its disposal, on its technical and human resources and, last but not least, on the funds which are available.

To forecast the wood supply, which can be expected from the country's forests, is as a rule, not so difficult a problem, if forestry is well established and no major changes in the forests are envisaged. Especially in countries with a long forestry tradition and in which long rotations are customary, the calculation of the future supply can be based on the fact that all stands, which will be harvested around the year 2000 or even 2030, do already exist. It is their development which has to be forecast.

Forecasting supply is more difficult, when a country is just starting with forestry or with the introduction of new methods or species, eg short rotation plantations of exotic species. There is in that case almost no data base available and very little or no experience to go by.

Things are difficult also, if old virgin forests are being replaced by second-growth stands so that the annual supply of wood is a gradually changing combination of old-growth and second-growth. However, these are technical problems, which good experts are able to solve.

The main factors to be considered in forecasts of consumption have been discussed at the beginning of this chapter. To recapitulate, they are: the development of gross domestic product, the availability and price of wood and wood products, the growth of population and the progress of technology both of the wood sector and of competing materials.

The forecasts of supply and demand will indicate what investments are necessary in the various branches of industry and in forestry.

Things get extremely uncertain, if imports of raw material have to be taken into account, because these depend on economic and political decisions of the round-wood exporting countries, which are, at least in the long run, unpredictable.

As a matter of course, but at the same time of great weight: decisions regarding forest industry investments should not take into account only the wood supply and wood consumption side, but also the very important question of labour force and skill, of energy costs and supplies, of transport and demand for forest products on the market and its possible development with regard to competitors.

3.4.3 Forest industries – objectives and options

Before considering the relations between forest policy and forest industries it may be helpful to look at the industries themselves. Various objectives may be pursued by the conversion of wood raw material into finished or semi-finished goods. In the market economies the prime objective of most forest industrial enterprises is to make a profit, in centrally planned economies to achieve specific production targets with the human, material and financial resources allocated for the purpose. In either case governments normally insist that these economic objectives be pursued in a way which is socially and environmentally acceptable; in some instances governments may give absolute priority to these social and environmental considerations either by legislation or by incentives. For example, the establishment of forest industries may be subsidized in some areas in order to create more jobs, or pulp industries may be forced to instal equipment to reduce the water and air pollution which they would otherwise cause. Another important government policy objective is to supply the population in the country with the forest products needed – furniture, houses, tools, paper etc. – at prices consumers can afford. A government may also wish to earn foreign currency by the export of forest products in which case a policy decision may be needed whether or not exports should have priority over the domestic market. In countries with scarce forest resources governments may seek to reduce the import bill for wood products by encouraging the substitution of wood by other materials which may be available locally, or by seeking to establish forest industries based on imported raw material as Japan has done. Even in countries rich in forest resources but with a large negative trade balance in the wood product sector – a not uncommon occurrence in developing countries – a government may be faced with a difficult choice of policy options. The annual cost of imports must be set against the capital cost of imported machinery for the establishment of forest industries and the fact that local circumstances may raise production costs above the price paid for the imports – technical and managerial expertise, energy costs, communications and other infra-structures are among the relevant factors.

Is small beautiful? This is another question posed by foresters, forest industrialists and governments almost everywhere in the world except perhaps North America, USSR and Scandinavia where there are very large areas of productive forest. Economies of scale in most forest industries are a fact, but they are connected with a high degree of mechanization or even automation, heavy capital investment and high output per employee and therefore fewer jobs than in smaller, less sophisticated wood processing plants. Large plants also require a larger supply of raw material resulting in higher transport costs where forests are scattered; they need a more highly skilled management and workforce and they may be less adaptable to changing market conditions than smaller plants. Not only in developing countries but also in developed countries with scattered forest resources, as in much of Western Europe, the economies of scale are in practice sometimes less than claimed. Where they are decisive, as in the manufacture of chemical pulp, a continued dependence on imports may be preferable to uneconomic small scale home production. Foresters nearly everywhere in the world attach a very high priority to the development of technologies which permit forest industrial plants and more especially pulpmills to be economically viable without being enormous.

Looking especially at the developing countries, the proportion of forest industries which have fallen far short of expectations is very high. What has gone wrong? There are difficulties of raw material supply, we find machinery and equipment not suitable for the purpose intended, inefficiency in management, various bureaucratic obstacles, such as the difficulty of obtaining permits to import spare parts and finally difficulties in marketing caused in part by the low quality of the product and in part by poor marketing. There are many possible causes for these disappointments, but a few basic lessons emerge.

1   A satisfactory raw material supply depends not only on its physical availability as revealed by forest inventories but also on the proper planning and organization of the harvesting, extraction and transport and on due account being taken of competing uses for the raw material, eg. for local crafts and domestic consumption;

2   The installation of unsuitable machinery is sometimes due to lack of appropriate technical advice but perhaps more often to lack of local knowledge by those who give the advice. It has also happened that what is needed is known, but something different is installed because that is the only equipment a donor agency offering help is able to provide;

3   There are no easy remedies for bureaucratic obstacles and inefficiency in management. A clearer definition of the objectives of management might help. There tends to be confusion between commercial objectives (making a profit or at least producing cost-effectively) and social objectives (meeting consumer needs or creating as many jobs as possible);

4   Perhaps the main lesson is that there is insufficient communication between the various parties involved: on the national side these normally include the ministries responsible for planning, finance and forestry and possibly also some state industrial corporations and private commercial interests; on the international side there are the various international and national aid agencies and credit institutions. Communication, however, is not enough. It is only when there is some national authority in effective overall control, both at the planning stage and, subsequently, to monitor implementation that there is a reasonable chance of success.

3.4.4 The relations between forestry and forest industry

Forest owners and forest industrialists have some interests in common and also some conflicting interests. Both benefit from a reliable flow of raw material from the forest to the industry and from a good market for forest products, but the forest owner wants a good price for what he sells and the purchaser wants to buy as cheaply as he can. These problems have been discussed under marketing and under forecasts. The forest owner also has an interest in the industry itself. In the

Figure 1 Tropical Moist Forest Selective cutting in a virgin forest in Indonesia. Because these forests consist of many species they are difficult to manage economically. Some are therefore converted to plantations, some are destroyed by shifting cultivation, others must be conserved because they constitute a uniquely rich and hitherto only partially explored gene bank. Photo. FAO.

110

Figure 2. Managed temperate forest. An uneven aged forest of several species on a private extate in Bavaria which has been developed by the owners (on the photograph) over a forty year period. This 'selection system' of silviculture requires highly skilled management, but creates forests which are very productive, beautiful and excellent both for soil conservation and as habitats for wildlife The system deserves to be encouraged where intensive multipurpose management is practicable. Photo: von Aretin.

Figure 3. Agroforestry Intercropping of *Grevillia robusta* with maize and beans. Nyambisundu, Rwanda. Modern agroforestry permits very intensive land use without detriment to the environment Copyright © Evans (author of Plantation Forestry in the Tropics (1982), Oxford University Press

Figure 4. Integrated land use A scene in Surrey, England An intimate mixture of human settlements, fields, hedgerows, copses and woodlands is typical of traditional European landscapes. The problem is how to reconcile the conservation of this pleasing landscape with modern farming and forestry Photo: Julia Hummel

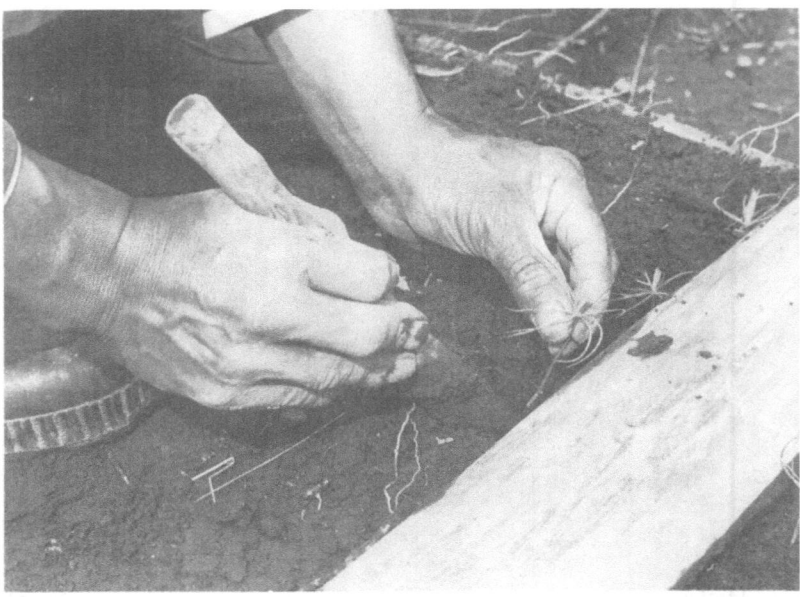

Figure 5. The start of the life of a tree. Transplanting a conifer seedling in a nursery in Guatemala. Photo: FAO.

Figure 6. The lifting and automatic bundling of transplants. Efficient forest nurseries are essential to the success of any afforestation programme. Photo: Forestry Commission, (UK).

Figure 7. Afforestation for sand dune fixation in Vietnam. The main object is to protect the agricultural land behind the dunes. This project was organized by the government with the help of the World Food Programme. A good example of 'Forestry for the people and by the people ' Photo: FAO

Figure 8. Multipurpose afforestation with *Pinus radiata* The main purpose in this case is erosion control, but the plantation will also provide useful timber. Photo. FAO

Figure 9. Afforestation in the United Kingdom. Large areas in parts of Western Europe are being afforestated in order to limit the region's increasing dependence on imports. Several million ha still await afforestation. Photo: Forestry Commission

Figure 10. A teak plantation in Honduras. A native of South East Asia teak like many other timber species is now planted far beyond its natural range of occurrence. Photo: FAO.

Figure 11. Mechanical Harvesting in a young plantation in Scotland. The cost of early thinning operations is high in relation to the price of the produce which mostly goes to be pulped or chipped. The development of improved technology, based on machines which are not too large is therefore vital. Photo: Forestry Commission (UK).

Figure 12 Log Transport Loaded truck on logging access road in California Timber is a bulky commodity, and an adequate system of roads is an indispensible prerequisite for economic forest management and development Photo: US Forest Service.

Figure 13. Shifting cultivation in Durango, Mexico, before roads provided access to the area. Without communications there was little scope for improving agricultural methods and for getting trees to a market; the trees were simply burnt to clear land for growing food in a very primitive way. The inevitable consequence is erosion and then more forest has to be cleared. Photo: Antony Hummel.

Figure 14. Erosion. This picture was taken near 13 on an area where similar shifting cultivation had been practiced a few years earlier. This area has now been opened up and programmes for rural development have been introduced which will alleviate the desperate poverty and reduce further erosion. Photo: Antony Hummel.

Figure 15. Shifting cultivation in Thailand Note the food crops between the charred stumps. Photo: FAO.

Figure 16. Forest Fire Damage in the USA Forest fires are among the major hazards against which forests have to be protected in many parts of the world. Many fires are caused by carelessness and the general public should be made aware of this Photo: US Forest Service.

118

Figure 17. Firewood – The poor peoples energy crisis. These women in Mali have to walk several miles to gather the firewood they need. Half the world's population depends on wood for fuel and several hundred million suffer hardship because there is not enough wood within easy reach. Relieving these shortages is a priority task in many countries. Photo: FAO.

Figure 18. Minor forest produce. Bee keeping in a pine plantation in Central America Apart from timber, forests are capable of supplying a very wide range of foods, medicinal products, fodder for cattle, fibres etc which must not be ignored when planning forest development Photo: FAO

Figure 19. Wildlife Management in Botswana. The maintenance of wildlife populations at a level which guarantees their survival but does not cause excessive damage to farms and forests requires sensible policies and the collaboration of government, farmers, foresters and hunters Photo: FAO.

120

Figure 20. National Forest Inventory in Mexico. A core is extracted from the tree on the left with an increment borer in order to determine the ring width from which the rate of diameter growth of the tree is determined. A national forest inventory provides some of the essential data for decisions on forest policy and at local level, for forest management. Photo: FAO.

Figure 21. Training. Students in Ecuador are being instructed by an FAO expert in the use of the chain saw for felling trees. The training of forest workers and their supervisors is essential for efficiency and safety. Photo: FAO.

Figure 22. Research. Controlled pollination of bagged cones Research and Tree-breeding in particular offers great scope for increasing the productivity of forest plantations in many parts of the world Photo: British Columbia Forest Service

Figure 23. Recreation. A patrol camp site in Deschutes National Park, Oregon, USA The provision of recreational opportunities is an important task of government forest services. Photo. US Forest Service.

Figure 24. Spreading forestry knowledge. A forester explaining forestry to a group of school children. Apathy and ignorance are among the forester's worst enemies. The cure starts best with the young. Photo: Forestry Commission (UK).

Figure 25. Inside modern sawmill. Note also the modern use of laminated wood beams in the structure. Photo: Western Softwood Ltd.

Figure 26. A small sawmill in a developing country. Photo: FAO.

Figure 27. Sawn timber yard laid out for side loader operator showing packaged timber stickered every row for air drying. Careful sorting, grading and handling are essential for economic working and marketing. Photo: Forestry Commission. Courtesy Western Softwood Ltd.

Figure 28. Forest Industry. The inside of a modern paper factory in Italy. The manufacture of paper has developed into a highly capital intensive and sophisticated branch of industry. Photo: FAO.

Figure 29. Forest Industry. A pulp and paper factory in Mexico with the township which largely owes its existence to the factory. Forest industries in addition to employing people create jobs in ancillarly services and amenities (shops, schools etc.) and can thus contribute to the general development of a locality. Photo: FAO.

first place he wants an efficient industry because it can afford to pay better prices than an inefficient industry; secondly, the forest owner wants to have within economic transport distance a whole range of industries in order to make the best use of his produce. A log suitable for manufacture into plywood will generally fetch a lower price if it has to be sold to a sawmill; sawlogs will fetch less if sold as pulpwood. There is also a national interest in having each log put to its most valuable use. Industrial efficiency is increased if the residues from one industry, e.g. sawmilling are used in another, e.g. for particle board manufacture. This type of horizontal integration can be achieved without the various industries being in the same ownership, although this may help.

Forest ownership by industries may affect other forest owners adversely because it strengthens the bargaining position of the industries in the purchase of wood. That is one reason why the acquisition of forests by forest industries has been limited in Finland and Sweden. In some countries forest owners' associations have themselves established forest industries or have been encouraged by government to do so. The vertical integration of forests and forest industry either upstream by industry into forestry or downstream by forest owners into industry facilitates overall planning and should therefore promote efficiency; this objective has been achieved in some cases; in others, vertical integration has been a dismal failure, for a variety of reasons but the root cause seems to be that good forest managers are not necessarily equally competent to manage a forest industry; and the same may apply in reverse. Governments should therefore be wary of promoting vertical integration as a matter of policy without at first carefully studying the likely consequences.

There are occasions when forest industries may try to persuade a government to direct owners to whom and for what purpose they should sell their produce. There was a case in point when the escalation of oil prices started to make it more profitable for forest owners in some places to sell wood as fuel instead of as pulpwood, with dire consequences for the pulpmills and for the balance of payments; it was calculated that twice as much oil could be bought with the proceeds of selling wood manufactured into pulp and paper than could be saved by burning the same amount of wood as fuel. Whatever the merits – and possible flaws – in this type of reasoning, governments should consider the following four points before deciding whether or not to direct supplies to particular industries:

1 Interference by government tends to be a disincentive to good forest management;

2 If it is necessary in the national interest to subsidize industries by deliveries of raw material below the market price, it is not the forest owners who should bear the cost but the nation as a whole;

3 Interference in a particular case may have wider, unexpected consequences;

4 Directions which cannot be effectively enforced are worse than useless.

126

Where a large concession is granted to a particular industry any private woodland owners within the catchment area of the concession also may have little choice but to sell their wood to that industry whatever the price that is offered.

Forestry necessitates long-term planning and this is greatly facilitated if a country has a clear policy concerning forest industrial development. This is particularly important in countries with limited forest resources where the establishment of a single major new pulpmill or sawmill may influence the wood processing sector as a whole. The development of a general strategy for forest industries presents difficulties. Government departments rarely have the technical expertise and the individual industries have too many vested interests to take an objective view. This is why an increasing number of countries are making use of firms of consultants which specialize in these matters. Clearly, where such consultants are employed, their findings should in the first place be used as a basis for informed debate by the various governmental and non-governmental interests concerned and not as an immediate blue print for action. Consultants too are fallible and may have vested interests.

References

Dieterich, V. (1953) Forstwirtschaftspolitik, Hamburg.

ECE/FAO (ETTS I) (1953) European Timber Trends and Prospects/1950-1960, Geneva.

ECE/FAO (ETTS II) (1964) European Timber Trends and Prospects. A new appraisal. 1950-1975, Geneva 1964.

ECE/FAO (ETTS III) (1976) European Timber Trends and Prospects. 1950-2000. Geneva. Supplement 3 to Volume XXIX of the Timber Bulletin for Europe.

ECE/FAO (1975) Forests and Timber: their role in the environment. Seminar, Interlaken, 1975. Madas-de Coulon Policy considerations.

ECE/FAO (1979) Joint Working Party on Forest Economics and Statistics: End-use Statistics. Geneva, 1979.

ECE/FAO (1981) Timber Bulletin for Europe. Geneva, 1981.

FAO (1967) Wood: World Trends and Prospects. Rome, 1967.

Goldstein I.S. (1979) Chemicals from Wood. UNASYLVA. Vol. 31, No. 125.

Leibundgut H. (1966) Die Waldglege. Paul Haupt, Bern.

Madas, A. (1974) World Consumption of Wood. Trends and Prognosis. Budapest, 1974.

Madas, A. (1971) Forest Sector Planning in Centrally Planned Economies, FAO/SIDA Seminar. FAO, Rome, 1971.

Mayer, H. (1977) Waldbau auf sziologisch-okologischer Grundlage. Fischer, Stuttgart & New York.

Poulsen, G. (1982) Non-wood Products of African Forests, UNASYLVA, Vol. 34, No. 137.

Speidel, G. and Steinlin, H. (1964) Moglichkeiten optimaler Betriebsgestaltung in der Forstwirtschaft, Bayerischer Landwirtschaftsverlag, Munchen.

SUBSECRETARIA FORESTAL Y DE LA FAUNA (1982) Vinculacion del Subsector Forestal con el Sistema, Alimentario Mexicano SAM SARH

THE BANK OF JAPAN, (1966) Hundred-Year Statistics of the Japanese Economy, 1966.

UN (1978) Yearbook of Industrial Statistics. New York, 1976.

US Forest Service (1974) The Outlook for Timber in the USA, Washington, DC.

US Forest Service (1975) The Nation's Renewable Resources, Washington, DC.

4   The service functions

András Madas

The service functions of forestry may be considered under two broad headings:

– the protection and, where practicable, the improvement of the environment (proctection against erosion by water and wind, conservation of wildlife and landscape etc.);
– the use of forests for recreation which is particularly important near towns and in rural areas frequented by tourists.

A major forest policy issue connected with these functions is the relevant costs and benefits which are extremely difficult to evaluate. The final section of this chapter is therefore dedicated to this issue.

4.1 Protection of the environment

4.1.1 The general situation

The environmentel role of forests can only be interpreted properly in the broader context of the human environment as a whole; land, air, water and all the plants and animals that inhabit our planet.

The relatively long and rapid economic expansion following World War II led to serious problems for the natural environment in the developed countries. Carson's 'Silent Spring' (1962) which showed in a dramatic – some would say exaggerated – way the consequences of an excessive use of pesticides, had a tremendous impact especially in the USA. For the great mass of people it became evident that unrestrained industrial development and the severe pollution associated with it was leading to environmentel degration which, unless halted and reversed, could lead to catastrophe.

In the developing countries the prime cause of environmental damage has been the population explosion which was brought about by better medical facilities and the successful combat of diseases such as malaria.

To feed the additional people, more and more forests had to be cleared for farming on land which rapidly became eroded under the primitive farming systems in use.

When it became evident that some environmental problems could only be solved by international co-operation a conference was convened at the invitation of the Swedish government in Stockholm in 1972. The principal document of the conference, the *Stockholm Declaration,* provides a summary of the basic environmental problems, which were also recalled by the VII World Forestry Congress in Buenos Aires, Argentina in the same year, 1972.

The 26 principles enunciated in the Stockholm Declaration include the following which are of particular relevance in the present context:

Principle 2: The natural resources of the earth, including air, water, land, flora and fauna and especially representative samples of natural ecosystems, must be safeguarded for the benefit of present and future generations through careful planning or management as appropriate.

Principle 3: The capacity of the earth to produce vital renewable resources must be maintained and, wherever practicable, restored or improved.

Principle 4: Man has a special responsibility to safeguard and wisely manage the heritage of wildlife and its habitats, which are now gravely imperilled by a combination of adverse factors. Nature conservation, including wildlife, must therefore receive importance in planning for economic development.'

The Declaration assigns a considerable role to the forest and calls for

- a full inventory of the forests of the world by the competent international organizations, using up to date technologies;
- large research programmes in the field of tropical forestry, land use, international trade of forest products and forestry legislation;
- dissemination of information to the participating countries about the prominent role of forests in erosion control, water management, protection of wild fauna and flora, preservation of sites essential to the tourist industry, etc.

The United Nations endorsed the Stockholm Declaration in 1972 and on the basis of the Declaration the Final Act of the Conference on European Security and Co-operation in Helsinki in 1975, amongst other things, agreed on co-operation in the field of environment protection.

A broad framework for considering the role of forestry in environmental protection was thus established.

The Stockholm conference was followed by a second conference in London in 1982 at which progress was reviewed. It was found that, although some progress had been made in the reduction of environmental pollution e.g. by the banning or control of some pesticides, on the whole the position had deteriorated; in the developed countries atmospheric pollution has increased and is thought to be responsible for the alarming death of fish in lakes and for the large scale dying of trees in parts of North America and Western Europe. In the developing countries, the problems associated with the clearance of tropical forest and erosion have got worse. Action by the governments concerned has become even more urgent.

Forests protect the soil against erosion by water and wind, they influence the water cycle and the atmosphere, they play an important part in the conservation of fauna and flora and they enrich the landscape. These aspects of the environmental functions will be considered in turn starting with the water cycle and the atmosphere, because it is the movement of water and air which determines the role of forests in the all important protection of soil against erosion.

4.1.2 The water cycle

Forests have a great role in the preservation and control of the water regime of a region. Water is drawn off the oceans by the energy of the sun in form of vapour, blown by the storms that circle over the continents, where it reaches the surface in form of precipitation to proceed in the rivers and then into the seas again, so maintaining an uninterrupted circulation.

Without water, life would be impossible for man, animals and plants; where man has interferred with nature the cycle may be disturbed, sometimes causing death and destruction. In some instances the interference may be simply caused by a population explosion. There may be enough water to support a small population and their animals and crops but not a large one. Deserts have been created in this way.

The estimated volume of water on earth is approximately 1400 million $\times 10^9$ tonnes. Over 97% of this volume is salt water; a further 2% is stored in the polar ice-caps, permanent snowfields and glaciers, and 0.6% as groundwater. Fresh water in lakes, the soil, rivers and the atmosphere accounts for only some 0.015% of the total volume.

Annual precipitation is estimated at $520,000 \times 10^9$ tonnes, balanced by an equal volume of evaporation. Components of the world's water balance are as follows:

		10^9 tonnes (1 e 1000 million tonnes)
Land	precipitation	108,400
	river run-off	37,300
	evaporation	71,100
Ocean	precipitation	411,600
	inflow of river water	37,300
	evaporation	448,900
World	precipitation	520,000
	evaporation	520,000

Source Encyclopaedia Britannica

The volumes are questionable, for example, estimates of annual run-off vary from 35000×10^9 to 45000×10^9 tonnes, but are adequate to give a general idea of the scale of the water cycle.

Except in certain localities, where water shortages are already a problem and could become worse, global availability of fresh water is expected to be adequate over the next few decades. What is of growing concern, however, is the quality of the water. Because of atmospheric pollution, the water that falls as rain or snow already has impurities in it. Chemicals used in agriculture and, on a very much smaller scale in forestry, as well as domestic and industrial effluents greatly add to the deterioration in quality. The forest's filtering capacity has a highly beneficial effect on that water which is precipitated onto and flows out of the forest.

National policies towards watershed management, therefore, including the use of chemicals in forestry, can have an important impact on the regularity of water supply to downstream countries, as well as its purity. However, positive efforts in these fields will have little value if they are not linked with and supported by policies by all countries using and discharging water into a given river system to combat water pollution.

Forests help to prevent floods. This is because where a watershed is covered by forests a smaller proportion of the precipitation flows to rivers in the form of runoff and infiltration than when a watershed is covered by grass. This is illustrated in Table 4.1. The beneficial effect of forests on flood control is, in fact, greater than is shown in the table because the runoff is much smaller than in other vegetation types and water which percolates into the ground reaches rivers slowly and therefore does not contribute to flooding. On the contrary, it helps to prevent rivers drying up when the weather is dry.

Forests slow down and regulate the surface runoff because a large share of the

Table 4.1 Water regime in areas with different plant cover (percent of total)

Plant cover	Distribution of precipitation			
	Evaporation from the crown of the tree	Evaporation before infiltration or runoff	Uptake by roots and net return to atmosphere by transpiration	Runoff and infiltration
Pure Spruce 60-80 years	30	2	57	11
Birch 40-50 years	15	7	54	24
Grass-land		73		27
Bare, intact soil		29		51

Source: Schretzenmayer et al.: Der Wald 1976

rain evaporates from the crown canopy without ever reaching the ground, another part running from the leaves and needles percolates into the loose, porous forest soil to be stored there and gradually released to move on downstream. The forest itself uses considerable amounts of water and returns it to the atmosphere through the transpiration process. Dense forests in the catchment areas of rivers therefore provide the best safeguard against floods.

There can, however, be situations where the complete use of the precipitation is justified for agricultural production.

Areas like this occur for example in California where the very dry climate led to the establisment of irrigated agriculture on a wide scale. In the upper reaches of such a catchment area, water-conserving coppice and shrubs are suitable; they make more of the precipitation available for irrigation than forests, while also providing effective erosion control.

Forests can replace artificial drainage by acting as enormous pumping systems able to extract daily by the energy of the sun large quantities of water from the soil and to evaporate it into the air. Polster (1981) found that in 40-50 year old stands the daily average transpiration in summer was as high as 53,000 litres per ha in douglas fir, equivalent to 5.3 mm of precipitation. Even in pine, where the transpiration was lowest it was 23,000 litres per ha per day, equivalent to 2.4 mm.

Another example is provided by the Forest Research Institute of Hessen (Federal Republic of Germany); this Institute made successful experiments with poplar plantations to drain a high plain with a clay soil, in which the water is not able to move laterally; for this reason drains could not extract enough water from the soil.

A special poplar species was planted which has both shallow and deep roots, so that the trees can both obtain the necessary oxygen and evaporate the desirable quantity of water; the soil was thus dried to the necessary degree. The advantages of this method are:

- the problem of draining the plain was solved;
- the method was relatively cheap;
- the necessary energy was provided by the sun;
- beside the reclamation of the soil, a considerable amount of wood was produced as a 'by-product'.

Another type of experiment was carried out in Hungary where, as well as the pumping ability of forests, the cleaning and filtering potential of forests has been successfully utilized.

In the town of Gyula two alternatives for a 'sewage' farm were considered. The first was, a traditional, artificial type of sewage farm, the second envisaged the planting of a 145 ha poplar plantation on a marginal agricultural saline soil, combined with a drainage system under the poplar plantation and a pre-purification installation.

The second alternative was accepted and it has been working with full success for a decade with the following technology. The sewage, after preliminary treatment, flows in open drains between the rows of the poplar plantation where the sewage infiltrates into the soil. The nutrient elements of the sewage are taken up by the poplar trees as well as a substanstial amount of water, which is transpired by the trees. The remaining part of the water infiltrates through to the drainage system, losing the rest of the polluting elements and taking with it some soda from the soil. The water leaving the drainage system has the prescribed quality. The advantages of this system are the following:

- it solves the problem of sewage purification of the town at one quarter of the costs compared with the traditional system;
- it utilizes the nutrient elements of the sewage and, in consequence of this the poplar plantation has a much higher annual growth rate than it would have without this treatment;
- the soil is gradually losing its salt content.

It may be worth mentioning that a big meat factory, which had previously erected its own sewage farm, decided to stop the operation of this sewage farm and joined the poplar system, because the operating costs were lower.

An earlier example from practice comes from Uganda where some 50 years ago several thousand ha of swamp around the city of Kampala were planted with eucalypts; as these use much water they dried out the swamps thereby eliminating breeding grounds for the mosquitos which carry malaria; at the same time, the plantations supplied large quantities of commercial fuelwood.

By the end of the century the supply of good drinking water will become a major problem in many regions. Artificial cleansing of water already consumed is not only costly, but reduces its quality for drinking. By the turn of the century water demand by industry, agriculture and for human consumption will hardly be met from the precipitation in densely populated areas and there will be an increasing need for water storage capacity. Forests can play their part in this.

For example, Pabst(1971) examined the role of forests in the water supply of the town Baden-Baden (Federal Republic of Germany) and found that the value of forests for this purpose amounts to at least twice their value for wood production in this region.

Pabst has used direct cost analyses in the following way: according to theoretical calculations, the water retention ability of forests in the area is 17% higher than on areas covered with grass; the difference is available for use as drinking water. The value is reckoned from the average precipitation, the area of forests and the price of drinking water in the locality.

4.1.3 The atmosphere

The complex relationships between forests and the atmosphere will be considered under five sub-headings; climate, the oxygen cycle, the carbon cycle, clean air and noise.

The *macroclimate* depends basically on the relative position and size of continents and oceans, on the pressure differences developed in the higher layers of the atmosphere, on the location of lowlands and high mountains. So it is hard to imagine that forests would have an immediate effect on macroclimate. There is no reliable evidence on this question, but there now appears to be a balance of evidence against the assumption that large-scale forest devastation changed the climate in the Mediterranean basis (Thirgood 1981).

A series of investigations and surveys have confirmed, however, the impact of forests on the *local climate* in and around forests. Schretzenmayer and collaborators (1976) demonstrate that about 5-6% of the precipitation in the Letzlinger Heide region is due to forests. The air moisture content is higher in the presence of forests, and interacts to create higher precipitation.

Atrokhin reports (1976) on the impact of forests on the volume of precipitation through the circulation of air masses above the forests, so giving rise to changes in air temperature, moisture content and air pressure. For example, he shows that in certain areas summer precipitation is 10% (35 mm) higher in forests than on land with scarce forest cover. If forest cover increases from 18% to 100%, precipitation rises by 60 mm. A 10% addition in the rate of afforestation results in a 2% increase in precipitation.

Factors such as precipitation, air moisture, temperature, radiation and wind in forests differ from those of open spaces. Transpiration raises air moisture in the forest. Depending on species, age of stands and on the season, the air moisture in forests can be 10-13% above open space values.

Comparing forests with surrounding open land the differences in air temperature are greater than those in precipitation and moisture content. Radiation is eliminated by reason of the uninterrupted overshadowing of the soil. Daily and yearly fluctuations in extreme values of temperature will thus be significantly reduced. This reduction in temperature variation considerably reduces the occurrence of early and late frosts.

Large scale investigations in *North America* revealed yearly average temperatures in the forest of 1.5-3.5 °C lower compared to the open air. Forests also slow down the melting of snow by as much as 14 days. Temperatures in a forest are generally cooler in summer, and warmer in winter than outside the forest.

Winds are highly influenced by forests. Depending on the height, species, exposure of the stand, the direction of the prevailing wind etc, forests reduce the speed and alter the direction of wind. Shelterbelts have a great impact on controlling the unfavourable influences of wind. They may reduce wind velocity in flat areas, facilitate the settling of snow for a distance of as much as 18-20 times the tree height, delay melting, reduce surface frost, control the soil moisture and micro-climate within belts, check wind and water erosion and drought, and add to average agricultural yield. (Atrokhin, 1976).

Gal (1972) found that shelterbelts may raise average yields of agricultural crops by up to 10% in the sheltered area; the increase in yield is sufficient to justify the cost of establishing the shelterbelts. Even larger increases in agricultural yields are reported from China by FAO (1982). Table 4.2 is reproduced from the above publication.

It was found that shelterbelts can increase agricultural productivity by extending the growing season for crops, by protecting seeds and seedlings from burial by sands and fine soil, and by allowing harvesting to be done in autumn without the wind shattering fruits and seeds. The favourable effects of shelterbelts are greater in areas of severe climatic conditions.

Of all harmful effects of wind, soil loss is the most significant. This can occur in certain periods on many kinds of soils, especially light sandy soils. Wind not only blows the soil away, but even bares the roots; moreover, the soil particles blown by the wind against plants, can injure them. Forests and properly established systems of shelterbelts are important elements in wind erosion control.

Soil erosion by wind is less visible than erosion by water and for that very reason perhaps more insidious. The dust clouds blown up by the wind on bare land when the weather is dry can of course be seen, but in the short term little change is often seen on the surface of the soil; in the longer term however there is a serious loss of fertility which is reflected in lower yields of agricultural crops. In arid regions desertification may be the final result. In order to reduce wind erosion, foresters should encourage the maintenance and planting of hedgerows and shelterbelts as windbreakss which slow down the wind, and thus reduce the loss and desiccation of the top soil..

Trees may also provide habitats for wildlife. The protective effects of shelterbelts generally extend, laterally, to about 10-15 times the height of the trees. The trees

Table 4.2 Effects of Schelterbelts

Effects of Shelterbelt	Chifeng Country, Liaoning Province, 4 rows, 8 m wide, 20 m high *Populus* spp.	Yu Country, Henan (Honan) Province, 1 row, 40 m apart, 20 m high *Paulownia* spp.
Wind speed reduction	58%	14-30%
Temperature reduction (spring and summer)	1° C	0.4-2.2° C
temperature increase (autumn and winter)	1° C	0.4-2.0° C
Evaporation reduction	38%	12-25%
Relative humidity increase	7%	13-20%
Grain yield increase	30-50%	13-17%

which in a comprehensive system of shelterbelts may take up about 10% of the total area will reduce correspondingly the area available for agriculture but this is usually more than offset by the higher agricultural yields and yield of timber from the shelterbelts. A difficulty arises from the fact that modern farming depends on the use of heavy machinery which can only be deployed economically when fields are large. A sensible balance must therefore be found between the economic use of such machinery and longer term considerations of conservation. Some farmers in developed countries are beginning to appreciate that it is in their own interest to stop the indiscriminate clearance of forests, small copses and hedgerows and even to plant new ones. In developing countries, where forest clearance is due to population pressures rather than the mechanization of farming, there is also a growing recognition for environmental forestry.

Interest has increased in recent times in the *oxygen producing ability of forests*. Up to now there has been sufficient oxygen both to cover man's own demands (burning up sugar to operate his muscles) and for the conversion of energy carriers (wood, coal, oil products, gas) into energy, but the rapid human population growth and the expansion of industry have now reached a point where a more detailed analysis of the oxygen balance is called for. This global problem is in fact receiving attention but no reliable figures or even suitable estimates are as yet scarce.

Let us consider first, the consumption, *Krebs* (1970) found that human beings consumed in 1970, 1,200 million tonnes of oxygen; in the year 2000 consumption could be expected to reach as much as 2,300 million tonnes.

The amount of oxygen necessary to convert energy carriers into energy is much higher. In 1980, 2,300 million tonnes of oxygen were taken from the atmosphere for this purpose and in return 30,000 million tonnes of carbon dioxide were released into the air.

According to recent forecasts of energy consumption, the following figures would apply by the year 2000.

Prospective consumption of oxygen

	1970	2000
	(thousand million tonnes)	
Direct human consumption	1	2
Technical consumption	14-18	45-50

There are significant differences between figures that have been published on oxygen production but the various authors agree that agricultural plants produce about 25% less oxygen than the forests, and the oceans three times as much as forests and agricultural plants combined. The range of estimates is as follows:

Estimated oxygen production of green plants (thousand million tonnes/year)

Resource	Minimum	Mean	Maximum
Forests	4.3	56	(66)
Agricultural crops	3.2	40	(49)
Oceans	22.5	144	(345)
Total	30.0	240	460

The estimated oxygen reserve in the atmosphere is of the order of $1,233 \times 10^{12}$ tonnes. This reserve may seem to be inexhaustible for the time being but this is far from certain.

One by one, developed industrial countries are already reaching the situation that oxygen consumption by mankind exceeds the amount produced by vegetation. Cole (1969) states that oxygen consumed in the *USA* is one and a half times as much as the quantity produced there by the green plant cover. Keeping life going under conditions of expanding industry is only possible because oxygen is replaced by the air masses coming from areas far from the USA.

More oxygen is consumed in *Switzerland* (*Krebs*, 1970) than is produced by forests and agricultural lands, yet millions of tourists are primarily attracted to that country by the fresh air as well as, of course, by the mountains, the peacefulness and good hotel services.

Scientists began to look around the earth to identify large oxygen producing areas. The Government of Brazil elaborated a large scale development programme for developing the *Amazon* basin to provide an impulse for agricultural and industrial production. Some experts reacted against this programme in an area containing the largest forest masses of the world and producing a volume of oxygen equivalent to 30-35% of the oxygen breathed by man. They felt that upsetting virgin forests might lead to serious damage to soils and the world's atmosphere. A world-wide demand emerged for safeguarding the oxygen producing function of forests. It has become clear that the problem of the oxygen cycle is not to be solved at an individual country level, but is of major international concern.

There are several possibilities and means to maintain the oxygen balance, but all these imply major shifts in technology from the burning of fossil fuels towards the more intensive use of atomic energy and of energy generated from water (tides etc.) and wind. Consumption of oxygen could be reduced also by replacing fossil fuels with electricity in the propulsion of cars.

The other approch is the extension of oxygen production itself. The trend in agriculture is to grow higher yields on an unchanged greeen area. Food demand trends are thus not being met by increasing the area under crops, but by raising average yields. However, no proportional increase in oxygen production can be expected.

Some experts believe that the world's major forests may have a role as oxygen suppliers depending on their composition and structure, while other scientists maintain that the forests' effect is neutral, i.e. that they consume about as much oxygen through the decay of organic material as they produce through photosynthesis.

The carbon cycle is, in a sense, the obverse of the oxygen cycle. Photosynthesis, the use of sunlight by plants as a catalyst for the conversion of water from the soil and carbon dioxide from the atmosphere into sugars and hence all organic matter, is the heart of the carbon cycle. Most of the world's carbon is stored in rocks. Carbonate sediments (limestones and chalk) are estimated to contain more than $10,000,000 \times 10^9$ tonnes of carbon; recoverable fossil fuels contain more than $5,000 \times 10^9$ tonnes (*Source*: UNEP, The State of the World Environment 1980; Chapter 2: Climate changes, deforestation, carbon dioxide and the carbon cycle).

Other major reservoirs of carbon are the atmosphere, with an estimated 700×10^9 tonnes of carbon, and the oceans, with some $36,000 \times 10^9$ tonnes. There is also the land biota and the soil. Life on land contains about 800×10^9 tonnes of carbon, of wich 90% is in forests, while the soil could contain between 1,000 and $3,000 \times 10^9$ tonnes of carbon.

Each year, photosynthesis withdraws from the atmosphere some $50\text{-}75 \times 10^9$ tonnes more carbon than is returned to the air by respiration. About 20×10^9 tonnes of this is taken up by forests. On the negative side, however, carbon is being returned to the atmosphere faster in total than vegetation is taking it up. The two main sources are the burning of fossil fuels and the loss of forest growing stock. It is estimated that before 1850, the concentration of CO_2 in the atmosphere was less than 290 parts per million by volume (ppmv). By the late 1950's the concentration had reached 313 ppmv and by 1978 330 ppmv. It is estimated that the annual input of CO_2 into the atmosphere is currently about 9×10^9 tonnes of which roughly half is from the burning of fossil fuels. The other half comes from burning of the forest and decomposition of organic matter.

These figures are only tentative estimates, except for those showing clearly the increase in the atmospheric concentration of CO_2. This increase is expected to continue at least until other sources of energy largely replace fossil fuels. Estimates are for the CO_2 concentration in the atmosphere to reach 365-380 ppmv by the turn of the century, and for the possibility of a doubling of the pre-industrial concentration of 290 ppmv within a hundred years.

The transfor of carbon from its storage in tree biomass to the atmosphere through destruction of the forest plays an important role in this trend, although how important is difficult to determine. One estimate indicates that the oxidation of the carbon contained in the standing volume on 8 million hectares of forest (a figure towards the lower end of the range of estimates of the annual loss of forest land) produces as much CO_2 as burning about 400 million tonnes of coal.

The main reason for concern about the increasing concentration of atmospheric CO_2 is the so-called 'greenhouse effect' – the warming of the lower atmosphere. While there is agreement that this could occur, however, there is uncertainty what the overall climatic effects will be. There could be various effects on the different snow and ice covers, for example, and the distribution and frequency of snowfall. It is remotely possible that a warming could melt some of the polar ice sheets, which would cause a world-wide rise in sea level.

The difficulty of predicting changes in climate as a result of the CO_2 phenomenon leads to the conclusion that it will be prudent to take whatever measures possible to avoid aggravating the situation, that is to say to reduce the net emission of CO_2 into the atmosphere. A policy to stabilize or increase the quantity of carbon stored in biomass would contribute to this objective. The most important step in this direction would be the conservation of the forest resource.

Forests have become increasingly important in the control and conservation of *clean air* because industries and motor traffic are polluting the atmosphere as never before.

Solid and gaseous particles in polluted air will reach the forests in different physical and chemical conditions. Grains of dust and suspended particles will be deposited on shrubs, trees, ground vegetation and the forests filter these out physically. The efficiency of filtration depends in part on the species composition, structure and height of the stand. There are still only a few data on the filtering capacity of the forests available. *Schretzenmayer* (1976) found that in a pure spruce stand 32 tonnes of dust per ha, in a Scots pine stand 36/ha and in a beech stand 68/ha could be deposited before reaching the physiological-ecological critical limit in needles and leaves. The next heavy rain washes the solid particles down, so needles and leaves become capable again of filtering.

Under central European climatic conditions the forest is an effective mechanical filter for particles from the polluted air. The large scale dying of forests since the late 1970's in central Europe and elsewhere which is believed to be caused in some hitherto not satisfactorily explained way by an interaction of these particles with SO_2 and NO_x in the atmosphere suggests that the limit of ecological tolerance is being surpassed on an alarming scale (see chapter 5.1).

Forests act also as an effective barrier against noise from traffic and other modern technology, which has become a grave environmental nuisance.

Various authors for example Keresztesi (1971) and Pabst (1971) show that forests, and plants generally, dampen noise by hindering sound wave transmission. Coming up against the forest, sound waves stop short, are reflected or are deadened. The intensity of the sound killing effect of forests depends on how dense the understory is. Evergreen coniferous forests are more effective in this respect than deciduous broadleaved forests. In a particular instance Pabst calculated that a forest belt 80-100 m wide would be as effective a sound barrier as artificial banks and would cost far less.

4.1.4 The soil

The prevention and halting of soil erosion by water and wind are among the prime environmental benefits from forests. There is evidence of this throughout the world. *Soil erosion by water* is especially severe in mountainous regions with high seasonal rainfall where forests have been cleared either for their timber or more commonly for farming. The soil gets washed away, farming becomes impossible, barren eroded slopes are all that is left; the farmer has to move on and the destruction continues elsewhere. Moreover, the damage is not confined to the slopes. Because of the more rapid runoff of water from eroded land, there is flooding in the valleys below after rains while during dry periods the rivers dry up. Where there are forests in the catchments the flow of water in the rivers below is more regular because forests act as a sponge, absorbing water when it rains and releasing it slowly afterwards. The damage done by forest clearance is aggravated by the fact that the soil and stones carried down the slopes with the water causes the silting up of rivers, lakes and reservoirs. In the USA, according to recent estimates, 30% of all existing dams will under present conditions, be totally filled up with sediments within 50 years and become useless. A further 24% of dams will cease to function for water storage in the following 50-100 years while only 38% will still be usable in a hundred years from now. More forests could at least retard the process.

Enormous damage caused by erosion can be observed in the *Dalmation Karst*, where the forests were felled and the entire region turned progressively to bare rock.

Looking at the deforested, stone deserts in the Near East, North Africa and parts of *Spain* it is hard to realize that these regions were at one time granaries for past civilizations.

Forests either on their own or, in certain circumstances in conjunction with engineering works are also a protection against avalanches which in some mountainous regions of the world claim many lives and cause much destruction each year.

Forests affect soil fertility in various ways, but more especially by the foliage which drops to the ground and then decomposes. In this way nutrients are recycled and the top layers of the soil enriched with organic material. This is generally highly beneficial. An exception may occur when certain tree species which are not suited to a particular site, are grown in monoculture. In that case the foliage returned to the soil will not decompose readily and soil fertility may suffer. The avoidance of unsuitable species, however is a matter of forest management rather than of forest policy. The use of forests and shelterbelts as a means of preventing erosion by wind were discussed in 4.1.3.

The main policy lessons to be learned from past experience throughout the world is that prevention is better than cure and that where damage has already occurred, the sooner action is taken the better.

Clearly, forestry to prevent soil erosion from water and wind must be considered not in isolation, but in the broader context of the present and future well being of those who live on and by the land. The productive management of forests and farms in a way that maintains soil fertility and prevents erosion is possible. This has been demonstrated both in developed and in developing countries, in temperate climates and in the tropics. Success depends on a sensible partnership between all who work and live on the land and the relevant local and central government authorities. It should be the policy of the forest authority of a country to play a leading part in the development of this partnership and programmes based on it.

Some large programmes have already been carried out; a few in Europe dating back to last century, such as the afforestation of the Landes in *France*.

More recently in *China*, over 30 million hectares of shelterbelts and forests have been planted to prevent water erosion and to fix the soil in the huge arid regions of the country. According to Gu Geping (1981) China in 1978 began planting a shelterbelt system which will ultimately cover more than 5.3 million hectares and run through northeast, northwest and north China. This gigantic project called 'the Green Great Wall of China', involves the participation of tens of millions of people. FAO (1982) reports that a single shelterbelt 1500 km long and 12 m wide was planted in only two seasons by 700,000 farmers from nearby communes.

In the USSR after World War II, a big plan had been set up to plant millions of hectares of shelterbelts in the lower Volga region where the dry winds combined with low precipitation caused regularly enormous damage to agriculture. The goal of establishing a huge set of belts was to brake the velocity of the wind and thereby preserve the moisture for the plants. According to statistical data over 11 million hectares of shelterbelts and forests had been planted in the Soviet Union by the end of the nineteen sixties outside the boundaries of the natural forests. Also in the *USA* several million ha have been afforested mainly in dry regions as a protection against wind erosion.

The 'green revolution' programme of the *Nigerian* government envisages a major system of shelterbelts combined with agroforestry and other measures in the north of the country in order to halt the southward spread of the Sahara desert, while at the same time providing the local inhabitants with wood for fuel and other domestic use. The desperate battle against the advancing desert also continues in parts of North Afirca, notably Algeria and Morocco. *South Korea* and the *Philippines* are other examples of countries which have embarked on major forestry programmes to serve environmental as well as the immediate needs of rural populations for food, wood and other produce of the land and also income from the sale of such produce.

4.1.5 Wildlife and landscape

The objectives are the protection of the still more or less intact part of nature, the safeguarding of flora, fauna and landscape against destructive pressures, the preservation of larger regions and historical sites for scientific purposes or for the purpose of education of the general public. Nature protection is a part of environmental activity, and should be without narrow economic concern and free from compromises. An important method of securing these objectives is by *the establishment and proper management of national parks* which generally include large forests; in these areas either all human activity is prohibited, or management is restricted to a definite level by regulations.

At the same time the importance of small scale conservation must not be overlooked. The hedgerows, copses, tree lined river banks even in densely populated areas are still a considerable reservoir of plants and animals. The economic gain associated with their destruction is often small, but the loss to wildlife and landscape great.

The first national parks were established in *North America* in 1872, and since that time a wide system of national parks has been developed both in the United Stated and in Canada. There is no forestry activity in these North American national parks. Efforts are being made to preserve nature in its original condition, above all for tourism and for scientific purposes. Preservation of the original conditions over the long term is hard to imagine without some human interference; for example, owing to large scale fellings in the adjacent areas and to ever increasing hunting activities, wildlife has tended to concentrate in the national parks, which offer an undisturbed refuge, but this can lead to an imbalance in the ecosystem. Overpopulation may be a concern, even in wildlife management, when there is no regulation either by man or by predators. Tourists in large numbers are also bound to influence plant and animal life in a national park and are incompatible with the concept of preserving nature in its original state.

Similar concerns apply to attempts to save the wildlife of *Africa, Asia and Latin America* from extinction. Established national parks provide protection for a considerable part of the high value wildlife. However, complete protection creates pressures on the ecological balance both in forests and elsewhere because with the extremely complex composition of wildlife the interactions will be altered, certain components will overreproduce, the balance of flora and fauna will be remodelled, forage will become scarce, and also other constituents will start to show signs of degradation.

Once the balance is disturbed thorough investigations are needed to establish the possibilities of achieving a new, dynamic equilibrium appropriate to the particular habitat.

In *Europe*, there are no large intact natural forests left, so the national parks and landscape protection areas have a different character to those mentioned above. Even most of the forests in the European nature protection areas have to be managed in order to preserve the original forest types or reconstruct them when they have become degraded.

The total interruption of human interference on large forest areas in Europe would generally result in their complete degradation and have an opposite effect to that desired. In the absence of their now extinct enemies such as bears and wolves, wildlife would multiply rapidly and prevent the regeneration of the forest which would inevitably become degraded.

An important objective in environmental protection is the *preservation of endangered species of flora and fauna*. Estimates of the number of species of fauna and flora inhabiting the world range from 3 to 10 million. A large majority of them constitute the forest ecosystem, especially the diverse, heterogeneous and complex tropical forest ecosystems. Extinction of species has continued for as long as life has existed on earth, when adaption has failed to keep pace with changes in the environment. However, until the present century, man's influence on the disappearance of species was small. That influence has grown immensely since the industrial revolution and the start of man's own population explosion. Loss of wildlife habitat may be the single most important factor involved, but changes in environment through pollution have also been having an insidious effect. All this is going on without the general public being properly aware of it, and there are only a limited number of field scientists who can really appreciate the extent of the risk. Extinction is an irreversible process through which the potential contributions of biological processes are lost forever.

The World Wildlife Fund has attempted to estimate, admittedly a rough exercise, the scale of extinction of species as a result of changes to the forest environment, notably loss of tropical forest. (*Source:* Special study undertaken by the World Wildlife Fund for the global report 2000 to the President (*op. cit.*)). It is believed that if present trends continue, hundreds of thousands of species could be lost by the year 2000, largely as a result of human intervention and on a scale that would make natural extinction trivial by comparison. Efforts to meet basic human needs could lead to the extinction of between one-fifth and one-seventh of all species over the next two decades, a substantial fraction of these losses being in the tropics.

All species might be necessary for future breeding; newly bred varietes – mostly in agriculture – outgrow the former breeds, and are introduced to replace them for commercial production. Shortages of the older varieties limit the gene pool still available for future selection, especially when breeding for resistance. Newly bred varieties are more productive, but at the same time they are, in a number of cases, more susceptible to pests and diseases. The future consequences of such breeding are impossible to assess.

Instituting gene pool centres and the organized preservation of still existing plant and animal varieties is a basic objective worldwide and a modest start has been made in forestry.

Some species are sophisticated indicators of changes in the ecosystem. Some threatened species – both of fauna and flora – are now rare, because during a long evolutionary period the living conditions for their existence had on the one hand been becoming less favourable for their reproduction, but on the other hand, had not become so bad that for the time being they were in danger of extinction. These species are now existing in biologically unstable conditions and in consequence of this, are much more sensitive to changes than other species. This is a valuable characteristic, which we can use in monitoring the changes in the ecosystem. This is done for example in Finland.

To sum up, the remaining natural forests constitute a genetic reservoir that man should, at all costs, seek to preserve over as wide an area as possible. The pressures on this reservoir are immense – growing populations, hunger, poverty – and the problems to halt the present trends are also immense. While some areas may require complete protection in many forests careful management and utilization may be compatible with conservation. It is of considerable significance that in quite recent times, the potential of the forest as a gene pool has come to be better understood by the general public, at least in the more 'educated' societes, and by politicians and decisionmakers everywhere. What has yet to be created is the political and legislative basis for coping with the problem, as well as the allocation of the very considerable resources that would be needed. Animal habitats may be preserved not only to prevent the extinction of species, but also for various scientific, educational and hunting reasons.

Game reserves and wilderness areas, both in industrialized and developing countries, have often been established with the objective of developing the tourist industry. Today safari holidays in East Africa have become popular with European tourists armed with foreign currency and cameras with telephoto lenses. The extensive areas of savannas, the natural habitat of lion, giraffe, elephant, zebra, wildebeast, and many other species, both predators and their prey, and the agreeable climate have proved ideal for international tourism.

In this particular case, it has not been necessary to make any major 'trade-offs' with other land uses, although the development of tourism may have to some extent affected the way of life of the indigenous population, both settled and nomadic. In other cases, for example the national parks and wilderness areas in North America, such trade-offs have often involved bitter political battles. The redwoods in California, for example, are famous throughout the world and the efforts of preservationists, including politically active groups such as the Sierra Club and the Audubon Society, have found support from outside California and the United States. At the local level, however, the transfer of redwood forests from timber harvesting to national parks has sometimes caused the closure of industries, loss of employment opportunities, and emptying of nearby towns and communities.

At the national level, the debate has been taken up by the forest industries, which have argued that the proposed withdrawal of extensive areas of commercial forest land in various parts of the United Stated could have serious implications for the national wood balance, the cost of forest products and the net trade position.

Another international aspect of wildlife refuges is migration, especially of birds. These creatures fill niches in the ecosystem at each end of their migration path, for example, insect control, and some, such as ducks and geese, are important for the hunting community as a source of food and revenue. International collaboration is needed, therefore, to ensure the conservation of the migratory species, including the protection of nesting sites and feeding places along the migration route. International pressure may also be needed to protect migrating small birds from wholesale slaughter by, for example, netting in certain Mediterranean countries. Among the various forms of vegetation, forests tend to be the most sophisticated and the strongest in influencing the landscape. This is the reason why forests are so vital to *landscape protection.*

We must be clear what we mean by landscape protection. The landscapes we love and seek to protect have been largely shaped by man. The pattern of woodlands, fields and human settlements which is so typical of many countries in the old world has evolved gradually over the centuries. Do we mean by protection the prevention of all further evolution of the rural scene or do we simply want to prevent the undue intrusion of modern technology – motorways, power lines, buildings out of keeping with the surroundings etc.? Most environmentalists would agree with the latter concept. The pursuit of an entirely static concept of landscape would be not only impracticable but also undesirable, but change to be acceptable must be gradual – most people are averse to rapid changes in landscape – and change must not lead to monotony. It is variety that enriches most landscapes; that is why the amalgamation of small fields into large ones, the elimination of hedgerows and small copses between fields and the conversion of mixed woodlands into plantations of a single species impoverish a landscape. A varied landscape is also a precondition for a varied fauna and flora.

Recently we can bear witness to a new wave of concern to protect landscape in many countries; the movement is strengthening its position in the parliaments also and wants to strengthen the regulations already in operation. Most of these, however, relate to specific, selected areas.

Let us take as example the Swedish Nature Conservancy Act, in which we find the following text:

'Landscape protection areas can be established in areas of outstanding natural beauty, e.g. places with fine views and tourist routes. Within such areas, it is not permitted, without a special licence from the County Administration, to build, or construct anything that will impair the beauty of the scenery. Licences are required to excavate gravel pits, stone quarries, etc. even outside landscape protection areas. The licence-holder is also required to restore the landscape to its former condition.'

Future generations will thank forest authorities which as a matter of policy pay due regard to the landscape in all forests, irrespective of whether there are any specific laws or regulations. Foresters can and should also play their part in arousing a general landscape consciousness where it does not already exist.

4.2 Recreation

4.2.1 Introduction

The use of forests for recreation became a major issue of forest policy in many densely populated developed countries as the mobility afforded by the motor car, increased leisure and more money enabled an ever increasing number of people to escape from towns to the countryside, where woodlands afford much pleasure to those who enjoy nature, fresh air and peace; woodlands suffer, as a rule, less risk of damage by visitors than farmland, because trees cannot be trodden underfoot like farm crops and, unlike cattle, they cannot escape through gates which are carelessly left open.

But also in many developing countries the recreational use of forests is becoming important for two reasons. First, green spaces in and near towns go some way to relieve the drabness and squalor of urban poverty; secondly, forests and their wildlife may attract tourists from abroad and thus help to earn foreign currency.

Numerous studies have been made in various countries in order to discover what type of forest people of differend ages, professions and social origins like best. Questionnaires, personal enquiries and other methods have been employed in these studies. Not unexpectedly, opinions differ widely, not only between countries but even within a country or a particular locality; and sometimes people even do not know what they would like because they hardly know what forests are and what they could offer.

There appears to be, among the many factors which influence personal preferences, a few on which there is a broad consensus of opinion:

– forests should not be too dense;
– there should be variety; different species with different coloured foliage or foliage which changes colour at certain seasons;
– there should be a mixture of tall trees, smaller trees and shrubs, open spaces to relieve the monotony etc;
– forests should blend in with the general landscape e.g. straight lines and sharp angles along the edges of plantations should be avoided;
– large, clear cut areas are disliked by visitors and opposed by the environmentalists and media.

The location of forests is also important. Forests in mountainous regions are reckoned to be more attractive than forests in plains. Lakes and rivers also increase the scenic value of forests, but to appreciate the scenery there must be openings in the forest or special view points.

The types of demand for recreation can be grouped as follows:

a General use of forests for recreation: – daily uses; – weekend and overnight uses; – recreation during a longer stay, generally during the annual holiday; – access for cultural and educational purposes.

b Special use of forests for recreational purposes: health, sport, hunting, fishing (angling), other uses.

As with other aspects of forest policy, a precondition for arriving at sensible decisions is a knowledge of the relevant facts and of the views and preferences of both forest owners and of the people who use forests for recreation. The smaller the forest area in relation to the number of visitors, the greater will be the need for careful planning and for appropriate legislation.

4.2.2 General recreation

Access on foot to publicly owned forests has been granted as a right or a privilege (where no legal right of access exists) in most countries. Access to private forests exists as a right in some countries but not in others. Forest policy as such, can have little influence on this issue, which is part of the much broader issue of the rights and duties associated with land ownership and occupation. In some countries, such as The Netherlands, the government offers financial inducements to private owners who permit access to their woodlands. Clearly, any policy concerning access must be supported by rules to define the duties and to safeguard the legitimate interests of the owner, the forest and the visitor. The rules should include, for example, precautions against fire, the exclusion of visitors from areas where wildlife needs to be left undisturbed or where visitors might interfere with or get hurt by logging operations.

Facilities. Most governments which, in response to rising demand, have embarked on a more active policy concerning recreation have wisely decided to concentrate on facilities such as signposted nature walks, information centres, campsites together with the necessary adjuncts such as car parks and lavatories. The advocates of dance halls in the forest have rightly received little support: those who like to spend their leisure surrounded by noise and crowds have plenty of opportunities elsewhere.It has also been generally accepted that the responsibility for providing leisure facilities should rest with the state and public authorities and that private woodland owners should have no obligations in this respect. In some instances, conflicts of interest have arisen between recreation and nature conservation. Visitors love wilderness areas and exploring wildlife habitats, but too many visitors destroy the unspoilt nature they want to enjoy. The problem arises mainly in densely populated regions, such as the countries of the EEC, where there are an average of eight people to every hectare of forest. In order to solve this, as well as other problems, some countries have pursued an active policy of trying to influence the demand by the so-called 'honey-pot' principle. Attractive facilities are provided in those parts of a forest which are least sensitive from the point of view of nature conservation. It has been found that only a relatively small number of visitors will go to areas without facilities or marked paths so that access need not be prohibited even in relatively sensitive areas. The Netherlands have been particularly successful with this policy.

For daily use there are the forests, parks, open woodlands, lakes and ponds close to cities or large settlements within easy reach by urban or suburban public traffic or private cars. The purpose here is to enable town dwellers after the work of the day, to take shorter or longer walks, to have a breath of fresh air outside the crowded and polluted urban environment, and to calm the nerves. The recreational use of forests on the outskirts of towns required the establishment of a system of well marked and maintained pathways, well located seats, rain shelters, occasional fire-places and, in the more frequented areas, 'keep fit' trails. Pathways should link up with urban traffic lines or come back again to the original place after a circular tour. In some places carparks may be provided adjacent to playgrounds for children and fire-places. In the more frequented areas the establishment of lavatories is required as well. Precautions are needed against the dumping of litter and arrangements made for clearing any litter that does get dropped.

Forests on the outskirts of towns are of special recreational value; public-ownership is therefore an advantage. Garbage dumping in these areas by lorries must be prohibited and effectively prevented.

Forests for *weekend visits* are generally rather farther from the larger towns, but within 100-150 km: they should be easily accessible by public transport, offer weekend relaxation and provide various facilities for entertainment and leisure. Larger centres must be developed for overnight stays, generally in the form of camp sites or rest houses. High standards of health and washing facilities are indispensable here, as well as at picnic places and other types of facilities for relaxation and entertainment. These can be installations such as 'keep fit' trails, larger open areas for different kinds of games including ball-games, interpretative trails for demonstrating forest trees, shrubs, simple forestry operations, shorter or longer hiking trails with vistas, possibilities for bathing, boating and fireplaces; anything, in other words, which contributes to a pleasant, not too noisy weekend.

Centres like this should be linked with the system of trails and established as hiking centres.

Care should be taken to allow for the possibility of separating people into small groups. Smaller car parks just off the highway may be appropriate to meet the needs of the family circle of not being disturbed by acquaintances or others. Small paths leading from these points for short sightseeing walks are welcomed.

Zones should be established in areas for daily or weekend tourism, so that particular needs may be met. Quiet and noisy areas should be marked off, as well as sites for physical training. Successful attempts of this kind have been made in The Netherlands and also in Hungary in the Pilis Park Forest near Budapest, which in 1981, was accepted as a unit in the UN MAB programme.

Hiking with cultural and educational interests has, as its objective, to make youth acquainted with the forest fauna and flora under varied scenery and field conditions. A series of camp sites may well aid this; but experienced guides are also needed to impart knowledge about the forest stands, wildlife, historical sites and folklore. Youth can in this way be persuaded to become involved with the protection of endangered birds. This educational experience can of course be combined with pleasure.

4.2.3 Specific recreational uses

Health requirements can often be fulfilled in the forest environment. For some illnesses, primary treatment is linked with stays in forests at different elevations. As an example may be mentioned here the formerly widespread disease in Europe, tuberculosis. Walks in peaceful surroundings for which forests are ideal are also a good precaution against modern occupational diseases, such as nervous breakdowns.

Forests with specific health functions may be located in the centre of larger forest areas in regions which are not downwind of industrial areas, and are preferably remote from population or tourist centres. Forestry operations in these forests around sanatoria must be subordinated to the special health functions.

There are numerous *sport* requirements. Groups of people taking part in organized, open-air sporting activities should be catered for as well as individual visitors and tourists.

For winter sports, sites for *skiing* may have to be developed in the forests by clearutting downhill lanes (pistes) and establishing the necessary facilities (ski-lifts, etc). Unless very carefully planned, these skiing facilities may, however, create grave erosion problems and increase the risk of avalanches.

Included here should also be forest *training grounds*, giving at the same time, depending on the degreee of advancement, opportunities for various physical culture activities.

Horse riding in forests has also become very popular in some countries but may lead to clashes of interest with visitors on foot. Riding may also add to the maintenance costs of forest tracks and paths. For these reasons in some forests separate tracks have been reserved for riders and for pedestrians and charges introduced for riding. Such charges, are, however, not easy to administer.

Hunting in forests has become a controversial issue in many parts of Europe. Because of the extinction of the main predators such as bears and wolves the stock of game rises to a level at which enormous damage is done to forests unless population levels of the game species are controlled by man through hunting. In countries where hunting remains the prerogative of a limited number of influential people game populations tend to be kept at a very high level so as to make it easier for hunters to obtain their trophies: under these conditions, the regeneration of forests whether by planting or by natural seeding, is only possible if the areas in question are protected against game by fences – an expensive operation the costs of which have normally to be borne by the woodland owner (whether public or private), and not by the hunters who are really responsible for the damage caused by the game. In other countries, on the onther hand, where every citizen can more or less shoot what he likes, populations of some species of game have been decimated, or even exterminated. A rational approach to hunting policies and practices could easily avoid these extremes but, unfortunately, when it comes to hunting, emotions and passions – some would say the instincts which man developed when his survival depended on his ability as a hunter – run high and stultify any attempt at a rational approach which would maintain populations of game at a level which forests can readily support.

Conflicts of interest may also arise between hunters and tourists, but these can generally be resolved by intelligent zoning and timing.

The important fact remains, however, that the view of hunters everywhere tends to differ from that of hikers, or even of foresters, who bear the responsibility for a harmonized realization of all the functions of the forest.

Hunting must be considered in the broader context of wildlife management. As policy objectives for this the EEC Commission (1979) suggested two main points which may also be applicable elsewhere. These objectives are:

1 To maintain a healthy, but not excessive population of as many species as are appropriate to a region and in harmony with local traditions;
2 To avoid as far as possible, interference with other aspects of forests management and agriculture, especially through game damage.

Hunting has been developed in some countries as an attraction for international tourists. Revenue from such activities, as well as the sale of skins, furs and meat on the international market can bring in a significant amount of foreign currency. National forest policies need to take this aspect also into account when seeking to establish a satisfactory modus vivendi between forest management and game management. One unattractive feature which has expanded in recent decades has been game poaching and the international trading of valuable skins and ivory, which more and more countries are making illegal. Efforts by countries to protect their wildlife against poachers must include provision of the forest service and other conservation agencies with adequate means and powers to do so, but will be useless unless strongly supported by legislation in consumer countries against importing and trading in the skins and tusks of endangered species. The same also applies to the trade in such animals themselves which may be sought by zoological gardens or as pets. Zoological gardens have, however, been playing an increasing role as refuges and breeding centres for endangered species. The European bison is one example and Przewalski's horse from Central Asia is another. In some developing countries, there is an additional consideration because rural populations may depend on game for food; in some instances up to 80% of the animal protein they consume comes from game. Here, the management of wildlife for food production should be a major component of land use policy and not only of forest policy.

Angling and fishing may have in some instances just as enthusiastic and large a circle of supporters as have other kinds of activities in the forest and cause far less controversy. Special enjoyment can be found in fishing in the ponds, lakes and streams in the forest.

The increasing inadequacy of water resources all over Europe is forcing governments to enlarge the capacities of water resources in the mountains and in forest areas. These can normally also be used for angling and fishing.

Forests are also used to provide other services unconnected with conservation or recreation. For example, forest areas are reserved for military training or for ammunition dumps.

4.3 Urban forestry

Much of what has been said about the conservation and recreation functions of forests applies also to what has become known as urban forestry, but the vicinity of a forest to a large town also creates special problems and opportunities: hence this separate section.

The following definition by Jorgensens (1970) of the scope and goals of urban forestry has become generally accepted:

> 'Urban forestry is a specialized branch of forestry and has, as its objective, the cultivation and management of trees and forests for their present and potential contributions to the physiological, sociological and economic well-being of urban society. These contributions include the overall ameliorating effect of trees on their environment as well as their recreational and general amenity value.'

Urban forestry is intermediate between horticulture and forestry and it requires an interdisciplinary approach. Landscape design, municipal water catchments, wildlife habitats, outdoor recreation must all be considered and the production of timber must not be ignored, because it may contribute to paying for the costs of management.

In densely populated parts of Europe the importance of city forests was recognized before the turn of the century and some of these forests were placed under the jurisdiction of the municipal authorities and multiple use management with emphasis on scenery and recreation introduced. In North America with its vast forests and low population density, urban forestry developed later. The very existence of vast forests may have been responsible for the fact that urban forestry continued to be neglected although more and more people lived in towns rather than in the country. But this has changed: urban forestry is now being taken very seriously in the USA and also in Canada interest has been awakened.

In the Soviet Union there always has been a very close relationship between man and forest and this has not been changed by urbanization and rapid economic development. The facts that there are very large areas of forest near cities and that 57% of the total land area is covered by forest may have contributed to this state of affairs.

Both in countries where the importance of the town forests has been recognized for a long time and in countries where this recognition has been more recent, there has been a switch of emphasis from traditional forest management by professional foresters to the multipurpose management called for by the special circumstances. Foresters are generally better qualified than others to assume overall responsibility but they must work in close collaboration with landscape architects, horticulturists, sociologists, town planners and others who may have something useful to contribute. Inevitably, lobbies with strong views but little knowledge wish to influence management and it requires political skill, tact and diplomacy mixed with firmness in addition to professional expertise to ensure good management and to protect these scarce green areas against the intrusion of urban developments.

The reasons for the increased interest in and political controversies over the management of town forests are the following:

- the area available for urban forestry is usually very small in relation to the number of inhabitants and of various interest groups;
- in many cities, the quality of the urban and suburban landscape has deteriorated in recent decades and the degree of atmospheric pollution and the noise from traffic have become virtually intolerable;
- improved standards of living and increased leisure enable more people in towns to take advantage of forests and other green spaces in the vicinity;
- land near towns valued at very little less than a generation ago may now be worth one hundred times as much. In qualitative terms, the increase in value of green spaces near towns may have risen even more.

The environmental and recreational services that trees and forests can provide for improving the quality of urban life are:

- influence on local climate. The temperature of the cities is generally higher than that of the parks and surrounding natural areas; so called 'heat-islands' are occurring in the cities. For example, in the city of Budapest the temperature of the Vermezo park is in the summer, 3-4 °C cooler than that of the streets of the city centre. In Montreal the summer temperature in the Lafontaine Park is 2-3 °C cooler than outside it. Forests, parks and green belts influence favourably, not only the temperature but also the humidity, wind and shade condition in the cities;
- trees and forests have a considerable screening effect and they reduce dust and pollution levels by absorption, diffusion, and reduction of wind velocity;
- noise abatement: In the previous sections of this chapter we have seen examples how forests act as a barrier against noise;
- provision of opportunities for active and passive recreation;
- outdoor education. In some urban forests, centres have been established where boys and girls can learn to recognize plants and animals, to take part in the protection of birds during nesting time etc. These activities have a positive influence on their education and future behaviour;
- beside the above mentioned benefits, the urban forests offer in some degree the general benefits and products associated with forestry.

From the point of view of forest policy, not the professional questions are the most important in connection with urban forests, but the political ones. This has implications for the training and recruitment of foresters for this type of work. Equally important is that the management of urban forests must be very closely integrated into the other aspects of community life.

4.4 The evaluation and financing of the service functions

4.4.1 Introduction

Many service functions of forests

- cannot be quantified in terms of money values; what, for example, is the value of preventing the extinction of some species of animal or plant or the conservation of a traditional feature of the landscape? Even where some quantification is possible as in the case of recreational facilities, the problem is usually very complicated;
- are not or only partially for the benefit of the forest owner but of third parties or

of society as a whole;
- create costs for the forest owner or others who produce the benefits (e.g. governments who pay grants to the owner).

The value of some forest products which are difficult to quantify must also be considered in this context. What, for example, is the extra value of firewood supply one km from a village instead of 5 km from it? For the women who, in many countries, have to carry the firewood, the benefit is enormous but cannot readily be translated into money terms.

Products and services that are not quantifiable in money terms are not taken sufficiently, if at all, into consideration in the planning and accounting systems of national economies. *In consequence the actual importance of forests is far more than the value shown in the statistical records of national income and gross domestic product.* It comes therefore as no surprise that *this undervaluation is also evident in the allocation of funds for forestry development all over the world.*

Because of the importance of these problems, the ECE Timber Committees and the FAO European Forestry Commission convened a meeting of experts in 1975 at Interlaken with the purpose of establishing methods to express the environmental and other service functions of forests in terms of value. At the meeting, a number of papers were discussed which analysed estimates of the social impacts of forests. The main point in the recommendations of this meeting was to improve statistical records on social and indirect benefits of forests as a basis for a more thorough analysis to determine the share of forestry in the national income. Other recommendations were that:

- larger amounts should be allotted in the future for the development of recreational services of forests;
- research work should be intensified into the whole range of forest products and services;
- on this basis the role of forests in the conservation and improvement of a healthy environment should be clarified.

Since that conference extensive research efforts have been made, but we are still far from the ideal of having an agreed objective evaluation system for the service functions of forestry that would serve as an objective tool of the decision maker. Too much still has to be left to subjective judgements and the influence of lobbies.

4.4.2 Methods of evaluation

Given this state of affairs those concerned with forest policy should have an idea of the methods which have been tried or suggested for evaluating the service functions of forests. They are as follows:

a   Analysis of the direct costs and benefits to the producer. This method is the nearest to the valuation using market prices. In hunting, for example, it may be possible to set the income from hunting against the cost of fencing and other measures to prevent damage to the forest from game; in the case of recreational facilities, although the benefits cannot be evaluated in market terms, the costs in materials and labour of providing these facilities are accessible.

b   Analysis of the profit foregone. When in the interest of the more intensive provision of social functions, timber utilization has to be restricted or made more costly, special management methods have to be introduced. In this case the profit foregone, or rather the difference in production costs, are considered as the value of the social impacts;

c   The method of comparative costs. This method gives information about the costs that become necessary, when, eg the service functions of the forest would have to be substituted by other – mostly technical – means. By this method the function of forests in erosion control, noise abatement etc. can be expressed in terms of value as already described in chapter 4.1.2.

d   Analysis of what beneficiaries would be prepared to pay. For example, the transport costs which visitors to forest recreation areas are prepared to incur give some indication of the value they attribute to such facilities.

One of the most comprehensive attempts to arrive at an overall evaluation of the service functions was made by Pabst (1971) in Baden-Württemberg, who used a combination of the above methods, to assess the costs and benefits of forests for water storage, erosion control, recreation, hunting, noise abatement etc.

The astonishing conclusion of this careful study was that the value of environmentel and recreational functions of the forest was many times more than that of the production in the densely populated and economically developed areas covered by the calculation.

These results cannot be generally applied to all regions, but give an idea of the orders of magnitude involved and show the elements to be used in establishing systems for analyses elsewhere.

Views and methods in North America – and in general in the Anglo-Saxon regions – are well known from the studies such as those made by 0'Connel (1975) and Grayson (1975) which are summarized below. Both studies were issued as documents at the Interlaken Symposium. The Anglo-Saxon valuation methods do not attempt to go into so minute details and precision as the German experts do. Approaches will be eleborated more in the light of commodity production and market laws.

O'Connel refers in his introduction to the fact that he is giving preference to economic tools and methods regardless of their evident deficiencies because they are much more usable than the alternative approaches: e.g. reacting to pressure groups or to personal bias. He points out that demand depends in general, not only on the price, but on the incomes, on the price of closely related alternative goods, on taste and prejudices. For recreation, the demand is in most respects similar to goods and services we generally buy. The main differences are the lack of market-determined prices, and the immobility of most resources used for recreational purposes.

He believes that the decision-maker, along with inputs from the public, must decide on the trade-offs between economic and non-economic concerns. This is an important assumption made in research; that people have two basic value systems, trade-offs in the economic area, and trade-offs in the non-economic area. To determine the final mix, open discussions are necessary between decision-makers and interested parties.

O'Connel reaches the conclusion that analysts renounce the use of economic tools to clarify this situation, but they emphasize that there is no common index available to measure all the pluses and minuses for a set of possible alternatives and he doubts whether there ever will be.

Crayson, based on experiences from the United Kingdom, highlights the importance of economic evaluation, 'so that too much has not to be left when taking decisions to the mercy of uninformed, but influential lobbies.' In his conclusion, Grayson suggests that case studies involving major attempts at evaluation of all costs and benefits can provide a most useful service by setting a framework within which informed debate can go on.

Das (1981) claims: that in India, the price realized by the sale of the wood from a tree may be as little as 0.3 per cent of its total worth if all the following factors are taken into account: oxygen production, control of air pollution and soil erosion, improvement of soil fertility, effect on water regime and humidity, shelter value to animals and birds. In some instances, there is also the value of flowers and fruit.

The proper evaluation of the social use of forests is a matter of interest also in the Soviet Union and other countries with planned economies.

Vasiliev (1972) reports that 34 million ha of forests in the Soviet Union are allocated solely to environmental protection and cultural functions, but the evaluation of the services provided by these forests is not easy. To avoid difficulties, he assumes that the values per hectare in industrial forests serving to produce commodities could be taken as a basic index of value for other forests as well. The impacts of shelter belts are evaluated by Senkevitsh (Atrokhin, 1975) using the following method.

Costs are calculated by adding the total – i.e. the direct and general – costs of establishing the shelter belt and the interest on the capital. For the valuation of the output, the increase in yield on sheltered agricultural lands that can be attributed exclusively to the effects of belts is taken into account as well as the income from timber production.

In this way, one of the environmental impacts of shelter belts can be transferred from the 'non measurable' products and services.

A different approach is necessary for the valuation of windbreaks alongside highways and railroads. Detailed analysis of the problem proved that windbreaks do not only protect roads agains snowdrifts, but diminish the risk of blowing out the permanent way on railroads and the adverse effects of wind on the speed of trains and on their fuel consumption.

Similar efforts may be observed in the other countries with planned economies, although here too there are conflicting opinions. Bludovsky (1975), the Czechoslovak author, for example, assumes that foresters in the countries with planned economies do not need to stress the social significance of forests and justify the necessary financial resources for their management by economic calculations. He argues that better forest management and systematic control provide an adequate security for the planned development of forestry and of favourable living standards. Suitable rules ensure that the uneconomic, but socially important jobs, get done and that illegal uses are prevented.

There are few who would agree with Bludovsky and we may conclude that numerical survey and economically accepted valuation of environment and social-recreational functions of forests are necessary both in countries with planned and with market economies. Independently of the social order, forests have a basically important role in the future of mankind. Global destruction of forests would make further development of human society impossible.

Only a few years ago, forests were still being considered basically as raw material resources, and the inventory of them and quantification in terms of materials, sizes and values could be easily arranged. But recently, the future of mankind as a whole became of central interest. This future depends first on the success of being able to avert the fatal danger of a nuclear war. Let us suppose and be confident that mankind can avoid this fatal danger. Then the next thing to face is the exponentially increasing demand for food and the more and more alarming pressures on, and pollution of, nature and the human environment. The impacts are not known exactly yet, even by the scientists and experts working in the special field. A worldwide monitoring survey, quantification and valuation is going on, but meanwhile *action must be taken on the basis of existing knowledge.* Decision-makers are only aided in their decisions, when the experts, scientists, specialists in the sectoral policies, determine to the highest degree all the interactions of forests with the economic development of society, with the improvement of living standards, with the social and health situation and with public attitudes.

4.4.3 Methods of financing

Finding the most appropriate methods of financing the service functions of forests is complicated by two difficulties: first, there is the difficulty already discussed that some of the benefits and costs can only be partially quantified in money terms or not at all. The second difficulty arises from the fact that some of the benefits accrue not to readily identifiable beneficiaries, but to a whole region or country or, in some instances, to humanity as a whole. Forests, in a water catchment area, help to prevent flooding further down the valley, irrespective of national boundaries. The conservation of the tropical forests in the Amazon basin may benefit the oxygen and carbon dioxide cycle throughout the world. Broadly, there are five main sources of finance:

- the woodland owner;
- the individual beneficiary;
- the state or a local authority;
- a non-governmental organization;
- international organizations.

The woodland owner, public or private, in some countries, is obliged by law to accept at his expense some constraints on forest management in the interests of environmental conservation or the provision of recreational opportunities. For example, clear fellings may be prohibited on steep slopes and visitors (usually only on foot) may have to be admitted free of charge. To the extent that owners of farms and other land are not subject to such constraints, there appears to be a good case for woodland owners to receive some sort of payment for what is a service to the public.

The individual beneficiaries can be made to pay if they can be readily identified as in the case of visitors to car parks or picnic sites with an entrance that cannot easily be by-passed. Even under these conditions, charging visitors is only worth while if there are sufficient visitors and the charges are sufficiently high to repay the cost of collecting the money. That is why, in some places, charges are made on public holidays when there are many visitors and entrance is free on other days. On the question whether or not beneficiaries prefer to be charged individually, Grayson (FAO/ECE 1975) is of this opinion:

> 'In matters of spending, people may well attach a lower value to a pound which the government takes to spend on their behalf than to a pound which they spend directly themselves. This is an awkward point, but one which cannot be ignored.'

The state and other public authorities generally pay for all costs arising in their forests and it is a common practice, in the interests of management efficiency, to identify the funds intended for specific environmental or recreational purposes. In the countries with centrally planned economies, this is the basic kind of financing of social functions. In several countries – e.g. in Hungary – funds for such targets are allotted on the national level and the relevant ministry provides the proper regional distribution.

Forestry undertakings, where environmental and social functions are of primary concern, should for practical reasons be incorporated either in part or as a whole into the budgetary financial system. It is nevertheless important in such cases to determine as exactly as possible the functional tasks, so that the budgetary system may not curtail, but even promote their realization.

For *privately* owned forests, government finance for the service functions can be provided in a variety of ways: by capital grants (e.g. for planting) annual grants (e.g. management grants), low interest loans, or indirectly by tax concessions. As individual service functions are difficult to evaluate, government finance is usually provided in recognition of these services as a whole, rather than for specific services but there are exceptions: as already mentioned, in The Netherlands the government pays a specific grant to woodland owners who admit visitors to their property; in Britain higher planting grants are paid for broadleaved species than for conifers which, in some parts of the country, are considered less beautiful in the landscape.

Non-governmental national organizations which contribute to environmental forestry have expanded their activities greatly in recent years, a reflection of the increasing awareness of the need for action, but their overall impact is still limited. Their activities are usually centred around the purchase and management of woodlands of special scenic or other conservational interest.

International sources of finance. International Governmental agencies such as FAO, UNDP and UNESCO, as well as non-governmental organizations, such as the World Wildlife Fund, have started to contribute to environmental forestry either directly, or indirectly. Some bilateral aid agencies are now also active in this field, but given the enormous international consequences of failing to act, and the fact that action is most urgent in some of the poorest countries of the world, there is a very strong case to expand these sources of finance for priority programmes, which are well conceived and likely to be cost-efficient.

References

Bludovsky, Z. (1975) Ekonomika viceučelové lesni hospodařstvi LESNICKA PRACE, Vol. 54, No. 10-11, Praha.

Carson, R. (1963) Silent Spring, Hamish Hamilton Ltd. London.

Cole, D. (1969) Umschau in Wissenschaft und Technik, Frankfurt.

Conference on European Security and Co-operation Final Act, 1975, Helsinki.

Das, T. (1981) The Value of Trees, Calcutta.

EEC Commission (1979) Forestry Policy in the European Community. Bulletin of the European Communities Supplement 3/79.

FAO (1972) Seventh World Forestry Congress, 1972. Unasylva, Special Issue, Rome.

FAO/ECE (1975) Forests and Timber; their role in the environment. Seminar, Interlaken, Switzerland. (a) Madas, A-M de Coulon: Policy considerations; (b) Atrokhin, V: Influence of forests on the environment; (c) Grayson, A: The relevance to forest policy of evaluation of environmental benefits; (d) O'Connel, P: Economic evaluation of non-market, goods and services.

FAO (1982) Forestry in China, FAO Forestry Paper 35.

Gál, J. (1972) Szélerózió-szélvédelem, Sopron.

Keresztesi, B. (1971) Magyar erdök, Budapest

Krebs, E. (1970) Wald und Luft. Allgeneine Forstzeitung, Wien.

Madas, A. (1978) Erdészeti politika, Budapest.

Molcsanov, A. (1973) Vlijanie lesa na okruzsajuscsuju szredu, Moskva

Pabst, H.R. (1971) Ansatze zur Bewertung der Sozialfunctionen des Waldes. Ministerium fur Ernahrung, Landwirtschaft, Weinbau und Forsten, Stuttgart.

Polster, H. (1967) Transpiration In H. Lyr, H. Poester u H.-I.Fiedler, Geholzphysiologie, Jena.

Que Geping et al. (1981) Environmental Management in China, Unasylva, Vol 33, No. 134

Savolainen et al. (1981) Scenic value of forest landscape, Acta Forestalia Fennica, Vol. 170.

Schretzenmayr et al. (1975) Der Wald, Leipzig.

Simpson-Lewis, W et al. (1979) Canada's Special Resource Lands: A National Perspective of Selected Land Uses, Ottawa

THE NATIONAL ENVIRONMENT PROTECTION BOARD (1972) Environment Protection in Sweden.

Thirgood, J.V. Man and the Mediterranean Forest, Academic Press Inc. (London) Ltd.

Thomasius et al (1978) Wald, Landeskultur und Gesellschaft, Jena.

UN (1972) Conference on the Human Environment, Stockholm.

Vasiljev et al. (1972) Lesnoje hozjasztvo v systeme planurujemoj ekonomiki, Warsaw

Vaux, H.J. (1982) Forestry's Hotseat: The Urban/Forest Interface, American Forests, May 1982.

5 Some special topics

Otto Eckmüllner and Adriaan van Maaren

This chapter is devoted to a number of topics which have either attracted particular political attention or which would otherwise have had to be split among several of the other chapters. The topics are: forest protection (against encroachment, disease, fire etc.), the concept of 'biomass', wood and energy, tropical moist forest, forestry in support of rural community development, agroforestry and farm forestry.

5.1 Forest protection

5.1.1 Introduction

Forests can only fulfil their various functions if they themselves are protected against destruction and damage. This protection raises issues of legislation, administration, management, research and international relations. These various measures and actions can only be really effective if they are co-ordinated and guided at policy level. Some of these issues are not only a matter of forest policy but also of other policies as we shall see. The three basic premises of a protection policy are:

a that forests are valuable and worth protecting;
b that forests are often endangered and need protection; and
c that the prevention of damage generally is easier and costs far less than curing the damage once it has occurred. Once forests have been destroyed or become badly degraded, it may not only be very expensive, but under certain circumstances quite impossible to restore them to anything approaching their original state, because forest destruction almost invariably leads to a deterioration of soil and site.

Forests are generally the more sensitive and vulnerable the more they have been influenced by man and are, consequently, farther from being natural forests with a good inherent power of resistence.

5.1.2 Past developments

For a very long time man did not appreciate that forests are necessary for the wellbeing of a country and its population and that they need, therefore, to be protected. The consequences of this attitude were in many cases very serious; millions of ha of good forests were lost and with the forests their production of goods and services, were, of course, lost too. Millions of ha of bare rock, of sand dunes, of steppe prevail now, where formerly forests had grown. Millions of ha of former good and valuable high forests have been degraded to poor scrub and bush forests like macchia or garigue.

Forest protection was only recognized as important and necessary, when timber shortages started to draw attention to the value and the economic importance of the forest. This was mainly the case where industry began to develop which depended on fuelwood (as for instance in the saltworks in Austria) or on charcoal (as for example in the iron industry in Sweden). One could almost state that in some countries the mining industry has been the 'father' of forest protection. Elsewhere, as in Scotland, where there was plenty of coal as an alternative to fuelwood, mining did not lead to forest management; most accessible forests were destroyed (admittedly mining was not the only cause) and then the mines switched to coal. Where mining for metals did lead to forest management there was a trend to convert mixed forests of broadleaved species and conifers into coniferous stands, often of a single species. The object was to raise production, but mono-cultures of conifers proved to be far more vulnerable. Thus forest protection developed into an important branch of forest practice and science; and also relevant legislation had to be introduced.

On a world view, the two most alarming developments in recent times are the continuing destruction of tropical moist forest estimated at between 5 million and 20 million ha per year (there are as yet no reliable statistics), and the recent spread of the gradual death of whole forests in parts of Western Europe and North America. These deaths are almost certainly associated in some way with atmospheric pollution, including 'acid' rain, but the exact causes have not yet been identified.

These two alarming trends illustrate the important fact that forests can no longer be protected by foresters alone; however, those responsible for forest policy have an important, indeed vital part to play. To leave such problems merely to others is no option because while to some extent the various sectors of society have common interests, there are also serious conflicts of interest which have to be resolved. For example the need for more land to feed rapidly rising

populations may be in conflict with the protection of forests against encroachment; sometimes the same people support contradictory policies: We all are in
favour of clean air but as producers or consumers of industrial goods many of us
are reluctant to pay the cost of preventing the pollution caused in their manufacture. Fortunately, some progress is being made towards getting the principle
accepted: 'the polluter must pay'.

5.1.3 Causes of damage

The main cause of all damage to forests is *man*. The damage he causes directly by
over exploitation, conversion of forests to other land use and by causing forest
fires is readily visible; but, as this sub-section will demonstrate, much of the
damage caused by pests and wind is also attributable to the way man has managed
forests.

Forest clearance

In many *developing countries* 'hunger for land' is the main cause of forest
clearance. There is a rapid growth of population, every year there are more
mouths to feed and more land is needed to do so; the only available land is usually
forest; thus millions of ha are cleared every year, the area is burned over and then
used to grow food and raise domestic animals. Most of the forest clearance is
unplanned and the methods of subsequent cultivation are primitive and lead to
erosion with the result that after a few years the area is lost to both agriculture and
forestry and the process is repeated elsewhere. Various lines of action are called
for to remedy or at least mitigate future damage:

1 the introduction of better agricultural methods which will raise productivity
   and at the same time reduce the risk of erosion. These possibilities are discussed further under agroforestry and forestry in support of rural community
   development;
2 the creation of rural employment outside agriculture; local industries, tourism
   etc.;
3 most important of all, the encouragement of birth-control without which all
   other measures can be no more than temporary expedients.

In developed countries forests are encroached upon by urban developments which
require land for housing, industries, roads, motorways, airports, power plants
etc. etc. These losses of forest area are often compensated by the afforestation of
marginal agricultural land, but usually, in remote areas or in the mountains. The
forest area may thus be in balance or even show an increase, but the development
as a whole may nevertheless be unfavourable, because the disadvantages of losing
forests near the population centres is by no means counterbalanced by the

advantages of creating new forests in places where the forest cover may be already sufficient and benefit fewer people.

It would be neither a practicable nor even a desirable objective of forest policy in any country to stop all forest clearance. What is, however, essential is to establish rules which will safeguard the legitimate interest of forestry. A danger arises from the fact that land sold for urban use generally fetches a very high price compared with its market value for forestry or agriculture. A forest owner is therefore easily tempted to sell his land for urban use or develop it himself for such purposes although the environmental benefits of the forest to all who live in the area may be immense. Unfortunately, it is not easy to quantify these benefits in money terms (see chapter 4). For these reasons, in some countries, the clearance of forests and their conversion to other land use requires the permission of the forest authority. The granting of permission is governed by guidelines authorized by government and enshrined in legislation. In Austria and Switzerland the forest authority may combine the permission with the condition that an adequate area be afforested elsewhere in the vicinity. The recipient may either undertake the afforestation himself or lay down the money so that the forest authority may do the job.

Austria has furthermore fixed in her new forest law of 1975, that 'Forest development plans' have to be worked out for the country as a whole using maps on a large scale of 1:50.000. These plans are a guide line for Forest Policy and for the forest authority. regarding the desirable future forest distribution, the necessary road-network to make the forests accessible, and regarding such matters as the desirable mixture of tree species. Of course, to decide or only to recommend, where forests should be and where they are not so necessary, is not a matter of Forest Policy alone; land-use planning needs a good co-operation with those responsible for agriculture, nature protection etc.

Over-exploitation

Another serious danger is the destruction or diminution of the forestry substance by practices such as overcutting, too big clearcuts that cause soil erosion, and 'creaming', that is felling the best and most valuable trees only, but damaging hereby the remaining stand or even destroying it. Commercial greed and corruption are generally the underlying motives although they may be cunningly concealed. In all these cases the 'plus' is taken by the present generation, the 'minus' is left to the following. Again it is legislation and supervision on the one side and information plus education on the other, which must be invoked if the forest substance is to be protected. There may also have to be rules on how timber is sold. Short term concessions in particular are liable to lead to over-exploitation and 'creaming'. Some countries prescribe the upper limit of the area of a cleacut, that limit often depending on terrain conditions: larger areas in harmless, flat terrain, but smaller areas on slopes. Re-afforestation within a specified time-span is also sometimes fixed by law. In some countries there is an obligation to manage the forest according to a forrest management plan, or age limits are fixed, below which tending measures only (thinnings for instance) are allowed, but no final harvest.

Whether hard, so-called dirigistic measures are necessary or a liberal attitude of law and supervision is possible depends very much on the forest owners' and the public's attitude and behaviour. A frequent error in Forest Policy is to try to solve problems simply by interdictions, by prohibiting measures, instead of really solving the problem in question; to work with such superficial measures will never be the right, successful way to solve a problem. It is therefore always necessary to look, where the very roots of the problem are! To prohibit for instance any cutting where people depend on the forest for their living, will never work. A sensible, controlled exploitation will be the better way, but most probably this alone will not suffice. What is needed in most places is a regional policy in which forestry and forest industry can play an important role. Overcutting and a poor state of the forests is often only a consequence of unsatisfactory, poor living conditions.

Damage by game
The damage that forests may suffer from excessive populations of game has already been discussed in chapter 4. The causes of these high populations are usually one or more of the following three:

1 the extinction of natural predators, eg bears and wolves in most of Europe:
2 hunters who encourage large game populations to further their sport;
3 a misguided concept of conservation; for example, it has happened that in a laudable desire to conserve certain species, wildlife reserves have been established, but no steps taken to prevent certain species from multiplying at the expense of others and of the forest. A large population of wild elephants for example can cause havoc to forests and, hence, to the habitat required by other animals.

The solution of these problems lies in a sensible co-operation between foresters, hunters, environmentalists and where wildlife is a tourist attraction, with tourist interests as well. It is easier to state this as a desirable policy than to implement it!

Damage by grazing
Grazing in forests – usually open forests with grass or other ground vegatation – causes various kinds of damage. Young trees are damaged by browsing; the trampling compacts the soil thus hindering aeration, drainage and regeneration of some tree species; on slopes grazing may encourage erosion; those responsible for the animals set fire to the forest because this may temporarily improve the growth of grass although it will eventually reduce fertility and cause erosion. Fires of

course may spread far beyond the areas to be grazed and cause untold damage.

In developed countries forest grazing has lost its former importance. Modern husbandry and cattle breeding need highly developed pasture techniques, whereas grazing in the forest offers almost everywhere a poor pasture only. However, the damages which happened in the past to the soil, the site and the forest itself are by no means overcome yet and even the best ways of restoring soil, site and stand are often not exactly known; too little research work has been carried out in this respect.

In many developing countries grazing in the forest is still very important; the most frequent animal is, in many cases, the goat, the most damaging 'Grazer'. It does not only graze on the ground; by browsing it damages and eventually kills the young trees at least up to man height. Millions of ha of formerly good forests in the Mediterranean Basin, in the Near and Middle East and elsewhere are more or less bare rocky land now and this was to a high degree due to the grazing of goats.

Grazing of sheep is also not so harmless, as is sometimes assumed, even if it is not so detrimental as goat grazing. Grazing sheep often cause soil damage and erosion, because on steep slopes they create small, horizontal paths and destroy the vegatation, which protects the soil. Moreover, sheep also damage and even kill young trees of some tree species by browsing.

The prohibition or restriction of grazing has on the whole not proved very effective, although there are exceptions. In Yugoslavia and Spain, for example goats were eliminated from large areas by government edict. In Cyprus at one time partial success was achieved by persuasion and education.

As a rule, however, a solution of all these difficult forest grazing problems can only be found by applying a whole set of policy measures which give the rural population a better living base by better farming as was done in much of Europe. The establishment of local industries, trades and handicrafts, tourism etc., to diminish the dependence on agriculture and grazing will also help.

Air pollution

The damage to forests associated with air pollution has reached alarming proportions in some parts of the world. The polluting substances include oxides of sulphur and nitrogen which cause the so-called 'acid rain' as well as a variety of substances such as lead, cadmium cobalt and fluorine which interact with each other and with the oxides. Most of the pollutants are emitted into the atmosphere by industry and traffic, but households may also contribute especially if they burn coal. The harmful effects have attained a measure which threatens *inter alia* even the future existence of forests. Some of the immediate effects on forests have been well known for many years and in a few instances forest owners have been paid compensation; that, however, does not help the forest. The longer term and seemingly much more widespread and serious effects are only partly known as yet, but are believed to be in some way responsible for the recent widespread deaths of trees and indeed of whole forests in parts of central Europe and the USA. These effects are not confined to the neighbourhood of the emissions; they transgress national frontiers and even continents and oceans. It is not only forests that suffer damage. The fish in lakes and rivers in parts of Scandinavia have virtually been exterminated as a result of the 'acid rain'. The pH of rain in these areas has decreased from around 5.6 in former times to 4.6 and is likely to decrease further to as low as 3.6 unless energetic measures are taken to halt pollution at its source. As far as forests are concerned, the effects are believed to be cumulative with little or no damage occuring until the critical threshold of resistance by the trees is reached. If this hypothesis is correct, the damage up to now is a warning of much worse damage to come.

These problems need interstate or even international solutions. Austria, for instance, in the heart of Europe, gets half of her sulphur load (of 38 kg annually per ha) from outside, from other countries.

Forest pollution on the ground
In many countries there is one more type of pollution which does not threaten the forest as such, but its functions; that is pollution of the forest with refuse. This forest pollution often begins in connection with outdoor recreation; wrapping paper, eggshells or orange peels are thrown carelessly away after a snack in the forest, that may lead to the dumping of empty cans and tins and eventually even of old broken down cars, old furniture, baths, in fact of everything, that people want to get rid of. All this is brought, often at night, into the forest and is left there. Everybody should dislike such an ugly mis-use of the forest, but it may be that it is at least to some extent, a fault of the municipalities, whether village or town, not to provide adequate facilities or, if there are any, not to tell the people, where they are and enforce their use.

In Austria, for instance, the fines for forest pollution can be very high – the equivalent of several months pay of a skilled worker – and most people would not risk to be fined, if they knew another way out.

Forest pollution is in many countries, becoming a serious problem, because it is not only ugly to see all this rubbish in the forest; it can and will most certainly lead in consequence to a pollution of the springs and of the ground water, whereas one of the most important services, forests render to the public, is to provide good, high-quality water.

Experience has shown that the best way of discouraging the dumping of refuse is to prevent it from starting. Many people hesitate to deposit rubbish unless there is already some present. That is why it pays to arrange for frequent litter disposal at any picnic sites or other places where there are many visitors.

Fire

Forest fires can be catastrophic, devastating huge areas of forest, destroying houses, villages and even killing people. Much depends on the climate and the type of forest. Conifers are generally, but not invariably, more susceptible than broadleaved species; and open forests with an undergrowth of grass, are more susceptible than closed forests. If the fire is a ground fire, trees with a thick bark may survive, whereas thin-barked or young trees or natural regeneration will perish. Older trees often get weakened only, but tend to be subsequently attacked by insects, which will kill them and then perhaps endanger other forests in the neighbourhood. A crown fire has always deadly consequences; even soil erosion may follow, when the protecting forest cover is lost.

Forest fires are – according to international statistics – to 80 or 90% caused by man, either by carelessness or, very often, by arson. In some parts of the world lightning is the main cause. To avoid forest fires started by human carelessness, information and education are the main instruments: these are matters that concern Forest Policy. Children should be instructed already in school, how dangerous it is to play with matches, and grown-up people should avoid smoking or burning an open fire in or near a forest in dry or hot weather. Carelessness can only be overcome by education, strict regulations and severe fines.

As far as forest fires caused by lightning are concerned, forest management can reduce the danger by a suitable choice of species, silvicultural systems etc. Large, even-aged stands of a single fire susceptible species should be avoided if possible. Very important is furthermore a good forest road network, so that any stand can be reached in the shortest possible time. The forest area should be subdivided by broad open strips as fire breaks, water tanks should be provided at the right places and natural water basins kept in order and full of water, and the necessary tools should be in place. Forest personnel must be trained and equipped to fight forest fires and where practicable arrangements should be made with local fire brigades, army units and any other organizations that could assist in emergencies. In some instances fire fighting arrangements can also usefully be co-ordinated with neighbouring countries.

Important too are the careful monitoring of fire danger (e.g. by weather indices) and good communications and public relations. Announcements over the radio and television and the closing of forests to visitors during periods of acute danger are among the measures that have proved useful.

Wind and snow

Wind and snow can be very dangerous and have often caused serious damage to the forests disrupting the market, because an oversupply of roundwood from stormfall or snowbreak may have a serious impact on wood prices. The damaged stands and treed have to be felled and removed as soon as possible, in order to avoid a deterioration of the wood and attacks of beetles and other insects, which could have far-reaching consequences. Resistance to wind and snow can be fostered by sound silviculture, especially an appropriate choice of species. In central Europe, for example, broadleaved stands and mixed stands of broadleaved and coniferous species have been found to be more wind and snow resistent than pure stands of conifers. Suitable thinning regimes have also been beneficial. Even more important has been the way fellings are conducted: stands must not be opened up on the side of the most dangerous wind – which is not necessarily identical with the prevailing wind.

Major damage by storm and snow impinges not only on forest management but also on policy because so many people and organizations must take co-ordinated action if the harm done is to be minimized: the necessary labour force and machinery may have to be procured from neighbouring districts; emergency roads may have to be built, precautions must be taken to prevent the outbreak and spread of pests.

Of great importance may be measures to prevent a breakdown in the market. These measures may include:

– organizing and supporting financially the storage of wind blown timber;
– facilitating its transport beyond the usual range of markets; even temporary exports, which might otherwise not be welcome, may have to be considered;
– forest owners in a whole region or country, may have to be persuaded to reduce their normal fellings in their own interest so as to prevent a collapse of prices.

These and any other measures that might be appropriate should be prepared well in advance along the lines of a military mobilization plan. It is too late to sit back and think what should be done and what should be done first when the catastrophe has already happened. Given proper contingency planning by the forest authority, in collaboration with others concerned, everything should run smoothly after 'pressing the button'.

Insects and fungi

Insects and pests are always present in any forest, they belong to the system. However in a healthy forest their number is limited and they are quite harmless. It is decisively important, not to offer these populations of insects or fungi an opportunity to 'explode'. Normally they live on sick or wounded trees or on broken down trees and it is their natural task, to work these trees up. The task of a good forest management however is to remove all accumulations of these dangerous breeding places and to burn them, if they are already invaded or infected, outside the forest, in order to destroy the next generation. Forest Policy has to take care, that appropriate regulations are fixed in the law and that the forest authority supervises the forests adequately. Sometimes the use of insecticides is unavoidable, in order to control the depredations of damaging insects. These means are sometimes toxic and should be throroughly checked and authorized by the authority concerned. There is a valuable development going on to produce means, which are effective, but almost harmless to the environment. It is worth noting that the use of pesticides in forestry never has been more than a minute fraction of their use in agriculture.

With regard to plant diseases every country will have to exert a certain control on all imports of plants or other material, which could carry dangerous insects or fungi. There exist already a number of international phytosanitary agreements concerning imports and exports of planting stock and timber.

5.1.4 General policy implications

It is a regrettable fact, that the world's forests are threatened to a high degree by man. Forest policy must therefore first of all seek to influence the mental attitude and behaviour of man towards the forest. One of the most necessary and effective means is *education*, but education of the population as a whole, beginning with the teachers of all types of schools, then from the teachers to the children at school and up to the adults, towndwellers as well as country people: by courses, information days, exhibitions, films, TV etc. The aim should be to incubate a careful and appreciative attitude in all towards the forest. Much harm to the forest happens out of ignorance, and this can best be overcome by information and education.

As mentioned in several places, forest potection must often start outside the forest: forest grazing ceases when there is better pasture outside.

Legislation is an important tool, but its value and its effectiveness depends very much on its contents being understood and supported by the public and on the authority, the 'forest authority' behind it, to enforce the law, including the fines. If a law remains 'a paper tiger', it is not worth more than the paper. Unless prohibitions are enforceable, alternatives should be chosen. In all these cases sound land-use planning is an indispensable prerequisit for an efficient forest protection.

In order to arrive at sensible protection policies, it is necessary first to investigate, what kind of dangers threaten the forest and where the roots of these dangers lie and then consider what can be done to prevent these dangers and to cure the damages, which have already happened.

5.2 The concept of biomass

5.2.1 Introduction

'Biomass' has become a topical issue since people have become aware of the finite nature of fossil fuels. 'Biomass' is a rather modern and wide term; it includes in its broadest sense all organic matter, vegetable or animal, alive or dead in any place, in fields, in forests, in stables of wherever. 'Forest biomass' is the same, but restricted to forest ecosystems. However, what is really meant here, is not the total forest biomass, but the 'phytomass' only and even not that in its full extent, but in fact the 'tree-biomass'. It should not be overlooked, that the trees of a forest are by no means the most numerous members of a forest ecosystem, though many people believe that a forest consists of trees only. However, in one hectare there are perhaps several hundred or a few thousand trees present, whereas the number of all members of the forestry biomass, down to the micro-organisms in the soil, is many millions or even thousands of millions per hectare. We are concerned only with the 'tree biomass'. The other elements of the forest phytomass – shrubs, ferns, herbs, grass, moss – are excluded from the discussion.

This tree-biomass consists of stemwood, branches, bark, stump and roots and of needles or leaves. The proportion of these parts is – within wide variations, depending on tree species and age – on an average:

60–65% stemwood
5– 8% bark
10–13% top and branches
10–15% stump and coarse roots
5– 7% needles or leaves
together 100%, all at dry weight

According to K.Kreutzer (1975, 1979) the average annual yield of a model stand of Norway Spruce (*Picea Abies*) on a first quality site with a rotation period of 80 years is, again in dry weight:

Stemwood (more than 7 cm ∅) without bark	4.290 kg/ha
Stemwood (more than 7 cm ∅) with bark	4.650 kg/ha
Branchwood with bark	718 kg/ha
Needles	380 kg/ha
Stump and coarse roots	1.370 kg/ha

For beech the corresponding figures are: 4.270kg/ha, 5.030 kg/ha, 897 kg/ha, 46 kg/ha, 1.253 kg/ha.

The reason why the forest tree-biomass in particular has received recently so much interest is, that wood is getting scarce, at least in some parts of the world, and that one possible way-out of this difficulty could be, to use in future a bigger part of this tree-biomass, than has been used in the past. Thus, the title of this section 5.2 'The Concept of Biomass' should be understood as 'Possibilities of using in future a bigger part of the forest tree-biomass'.

The answer to this question cannot be a simple 'why not?' In fact, the problem is a very complicated one and includes not only the question, what could be used more than in the past and what could this plus be used for, including the problems of costs and returns in harvesting, transport and utilization; there is also the question to be answered, what the reaction of the forest and of the forest soil and site and, last but not least, the impact on the further production of wood and services will be. It would be rather careless and irresponsible to start with using a bigger part of the forest tree-biomass without knowing the outcome.

5.2.2 History

In many parts of the world, where already centuries ago, wood had become a valuable raw material for a host of purposes, as a rule the stemwood only has been harvested and removed. This means, however, that 35-40% of the tree-biomass remained unused in the forest: bark, branches, stump and roots, needles or leaves. These parts of the tree-biomass were, however, not in the least useless or lost; they were transformed to humus and mixed with the mineral soil. The nutrients, which they contain in a considerable measure – compared to wood – were set free and put at the disposal of the forest again. This is the so-called 'nutrient-cycle', the self-fertilizing system in forestry, which is extremely important and on which depends the yield of the forest, the yield of wood, but also of services. Thus, the forest's soils and sites did not suffer by the harvest when stemwood without bark only was extracted; the utilization of the forest in this way could go on and on without any noticeable reduction, because the little amount of nutrients contained in the wood – wood consists, to express it in a very simple manner, in the main of air (CO_2), water and energy (sunshine) and of a very small amount of nutrients taken from the soil – is normally easily replaced by the decomposition of the subsoil and by the influx of airborne nutrients precipitated as solid particles (dust) or in rain.

In some parts of the world, however, man did not restrict his utilization of the forest to the stemwood only. He used the whole tree--biomass, the stemwood, top and lop, bark, stump and bigger roots; and even the needles and leaves from the trees, and those lying on the ground as forest litter from former years, were collected and used, the latter as bedding material for domestic animals. This has in fact been the full use of the tree-biomass. The reason has been, that too many people had to live with their animals from too small agricultural areas, which had to be fertilized at the cost of the forest, if their fertility was to be maintained.

Since World War II in the *developed countries* the methods of agriculture have been much improved and the full use of the forest tree-biomass became unnecessary and has been completely abandoned. However, the forests have not yet recovered from the damage to soil structure and the loss of nutrients brought about by the extraction of the full tree biomass. The forests concerned are poor, mostly secondary pine, slowly growing thus producing very little wood and, as a rule, producing very few services too.

In *developing countries*, especially if they are densely populated or have a marked increase in population and only few and poor forests, the harvest and utilization of a big part of the forest tree-biomass was and is widespread, mainly in order to cover the needs of energy, of heat for cooking and heating. This problem will be dealt with in more detail in chapter 5.3 'Wood and Energy' and 5.6 'Agroforestry'.

5.2.3 Policy aspects

While in former times the necessity of using a big part or even all of the forest tree-biomass had been caused in the industrial countries of today by the then under-developed state of agriculture and by structural deficiencies, mainly too small farm properties, now the reasons would be quite different: lack or scarcity of wood raw material mainly for industrial processing and the search for renewable sources of energy to replace fossil fuels. In the developed countries, the high standard of living has led to a very high consumption of wood – in buildings, in furniture, in newspapers and books and so on. Forest industries have enormously developed and their demand for wood, for industrial wood of course, can hardly be covered any more. 'Under-supply' of forest industries is, thus, the main problem in these countries now, although the possibilities of using more wood biomass for energy and for other purposes (e.g. cattle fodder by fermentation processes) are also being actively explored.

First of all it is necessary to investigate whether this scarcity of wood is really a fact or if for instance the present use of wood is perhaps unduly big and even wasteful; or whether better processing techniques or economizing measures of various kinds could help to ease the situation. It has to be asked, whether using a bigger part of the biomass is really the only and best way, the most appropriate way to solve this problem.

Are there no other possibilities to increase the availability of wood? What about promoting, for example, the use of wood residues, first of all those which are generated by forest industries (sawnmills for instance) or the use of waste paper or the re-use of wood, which has already been used once (construction timber from repair works of buildings or old crossties of the railways etc.) or could forestry itself increase its production and productivity, for example by enlarging the forest area, by introducing genetically improved planting stock, by planting fast growing species like poplar, willow or eucalypt? Existing deficiencies in the forests like poorly stocked or degraded stands, stands of unsuited species etc. could and should be remedied, and the total forest area made fully productive except where there are specific environmental constraints (prevention of erosion etc.).

It seems, furthermore, that in an increasing number of countries the forests are under-utilized, that the growing stock is, therefore, increasing every year, not always to the advantage forest and of the economy.

The utilization of a bigger part of the forest tree-biomass is thus one possibility only among a number of others and it should well be considered, which one should be preferred. An important question may be, inter alia, if such a bigger use would not have unwanted or even dangerous effects on the function of the forest, on the productive function as well as on the service function. Often all these functions are important and valuable and none of them should be injured by a 'new concept'. If, for example, the stumps and major roots were extracted, how would the forest soil react? Would not its structure be heavily disturbed or even destroyed, would not soil erosion follow? Or, if fertilizers are applied, in order to compensate for the loss of nutrients would not springs and groundwater suffer?

All these pros and cons should be well considered and weighed. The risk to come to an, at least in the long run, wrong decision should not be overlooked. The additional raw material, which could be extracted from the forest, is at any rate of low quality and value and the losses and damages to the forest, its soil and site and of the environment are perhaps much bigger.

As a rule, forest soils are not the best ones in a country; they are the 'left-overs' of agriculture, which occupied the better soils already long ago. It is therefore all the more important to maintain the state of the forest soils or even to improve it by good silviculture and careful treatment. Whether the stemwood without bark only or more of the tree-biomass is extracted, makes a big difference; the following table illustrates this difference for Spruce (*Picea Abies*) first class site, rotation 80 years, according to K. Kreutzer (1979)

Loss of nutrients in kg/ha during one rotation (80 years)

Kind of harvest	N	P	K	Ca	Mg
1 Stemwood timber (7 cm $\emptyset$ +) without bark	290	18	180	310	56
2 Stemwood timber in bark	480	42	300	570	76
3 The same, but with branches in bark	1490	155	760	870	165
4 The same plus stumps and course roots	1570	163	870	1020	180

Thus, the difference in the loss of nutrients can be three – to fivefold! Some of these losses are replaced by nutrient input from rain which contains nitrogen and from the influx of dust and aerosols. Thus, the nutrient-balance looks somewhat better (again according to K. Kreutzer 1979)

Average annual nutrient-balance for spruce in kg/ha over rotation

Kind of harvest	N	P	K	Ca	Mg
1 Rough timber (7 cm $\varnothing$ +)					
without bark	+11,9	+0,14	−1,8	−4,1	−0,90
2 Rough timber in bark	+9,5	−0,16	−3,2	−7,6	−1,15
3 The same plus branches in bark	−3,1	−1,57	−9,5	−11,4	−2,26
4 The same plus stumps and					
course roots	−4,1	−1,67	−10,2	−13,3	−2,45

As can be seen from both tables, the losses and the nutrient-balance differ widely, whether wood without bark only is removed or more is removed. The final consequences on soil and site will depend decisively on the forest soil and its reserves of nutrients and on the input by the decomposition of the subsoil; on the lower parts of a slope the nutrients washed down from higher up may also affect the issue.

European experience over a very long period has demonstrated the dangers of removing too large a proportion of the biomass and data such as those quoted above explain some of the likely reasons. Generally speaking very little harm is done if logs are removed from the forest with their bark as is now generally the custom. The removal of branches and lop and top is more serious, but not as serious as the removal of foliage which has a higher nutrient content. The worst damage may result from the removal of stumps with roots because of the effect on soil structure and drainage. Fortunately, it is rarely economic to utilize small branches, foliage and roots.

Direct intervention to prevent the exaggerated removal of biomass for industrial purposes is likely to be necessary only in exceptional circumstances because the cost of removal is high in relation to the value of the produce. In subsistence situations, the situation is quite different, because people may depend on this biomass for domestic fuel.

In either case, forest policy's main task in relation to biomass does not lie in the adoption of any extreme new measures but rather in the general strengthening of measures which safeguard the forests and promote their production potential and its utilization both in the short and in the long term. As already mentioned, the reduction of waste in processing and the recycling of wood can also make a significant contribution.

Where the use of more than the stem of the trees cannot be avoided the likely damage will be less if at least the roots and foliage are left and if the practice is restricted to level ground and to better soils rich in nutrients. On poor stoney or sandy sites with a low nutrient content any use of more than the stemwood is particularly damaging.

Additional forest policy objectives should be to ensure the collaboration of foresters with agriculturists and others concerned with the uses of biomass; and to promote relevant further research and investigation.

5.3 Wood and energy

5.3.1 Introduction

The energy problem of today or, as it is often called 'the energy crisis of today' has two quite different aspects:

In one part of the world, in the bigger one, where about two thirds of the population live, that is in the developing countries, there is a growing scarcity of energy, but of energy for simple daily use: for cooking and heating! The main source of this energy is wood and where this is not readily available various plant residues and dung. In spite of a very or even extremely low per capita consumption of half a m^3 of wood per year of less, the energy scarcity develops sharply. The main reason of the energy problem is the increasing imbalance between the number of people, which is ever increasing at a rate of 2.5% or more per annum, and the decreasing supply of fuelwood from the forests. The forests are, as a rule not managed, over-exploited or even depleted and destroyed. In a number of countries the development is alarming. Effective policy measures are urgently needed.

In the other part of the world, in the industrialized, developed countries, energy consumption in total and per capita, is very high but the population is more or less stable or even slightly declining. The main consumers of energy are industry and traffic, but households, too, play a considerable role. The sources of energy in these countries are in the main oil, hydro-electric or atomic power and coal, but the costs of all these kinds of energy exploded during the 1970's. The problem is, therefore, first of all to reduce the consumption of energy or, at least, to slow down its increase and to substitute it by less costly kinds of energy wherever possible. The consumption of energy is linked with economic growth in these countries and, therefore, with the standard of living; at the same time the increased consumption of certain forms of energy (oil and coal), has added to the pollution of the atmosphere and thus to a worsening of the 'quality of life'.

Wood and energy problems of the developing and industrialized world need to be dealt with separately. The roots of the problem are different and the ways and means to solve them or at least to ease them are different too.

5.3.2 Some facts and figures

Fuelwood and charcoal and, to a lesser degree water power and windmills too, have for thousands of years, been the main or only source of energy for mankind. It should, however, be noted, that fuelwood and charcoal are essentially stored solar energy. Nature has this wonderful means to store solar energy simply by producing wood or other plant material; even all fossil fuels like coal, oil or peat which are a result of former photosynthesis, are also stored solar energy.

Regarding the question 'fuelwood or charcoal', which is rather important in the *developing countries*, it should be noted, that in the process of charcoal production about one fifth or even one third of the energy of the wood is consumed and, thus lost. On the other hand, charcoal has certain advantages, in transport for instance, because it is much lighter and less bulky, and in use because it is easier to handle and causes less smoke. Charcoal is therefore preferred in urban centres, but not seldom one third of the family income has to be spent for this purpose.

The energy consumption of mankind was very modest as long as the population was small in number and lived from hunting and collecting fruit or later on from agriculture. This is still the case in a few developing countries; but in most, a population increase, which is partly due to the great progress in medical care, led to a much bigger consumption of fuelwood and charcoal and this, together with the increasing use of wood for housing and other local and domestic purposes, resulted in a heavy over-exploitation of the forest which is often a not very productive 'open forest', or savannah forest. These developments were aggravated by uncontrolled forest grazing. A part of the forest area has, moreover, been cleared for agricultural or other purposes and the consequently smaller, over-exploited forest area could no longer cover the increasing needs. The problem is the more difficult, as about 80% of the total energy consumption has to come from the forest.

In these countries the energy situation has become often an extremely serious problem. There are more than 100 million people, who suffer already at present, under an 'acute scarcity', which means, that their fuelwood sources are depleted and that their needs cannot be covered, even not by over-exploitation, because there is nothing left to exploit. There are often only 0.05 or 0.10 cbm fuelwood per capita and year available, whereas the needs would be five or ten times higher.

According to FAO (1981) there live in 'acute scarcity'

in Africa	55 m. people, of which 49 m. in rural areas
in Asia Pacific	31 m. people, of which 29 m. in rural areas
in Latin America	26 m. people, of which 18 m. in rural areas
Total	112 m. people, of which 96 m. in rural areas.

Many more people live under 'deficit conditions', by which is meant, that minimum needs can only be covered by over-exploitation, which leads, of course, to a progressive depletion of the forest resource. Under these 'deficit conditions' live:

in Africa	146 m. people, of which 131 m. in rural areas
in Asia Pacific	832 m. people, of which 710 m. in rural areas
in Latin America	201 m. people, of which 143 m. in rural areas
Total	1283 m. people, of which 1052 m. in rural areas.

It is extimated that the number of people living under 'acute scarcity' or under 'deficit conditions' will increase by the end of the century to almost 3000 mio, of which 2400 mio are expected to live in rural areas. Thus, the problem is extremely serious, exremely big and extremely urgent. The 'energy crisis of the poor man' is already present, but will get worse in future, in the near future.

In a paper, which had been submitted to an expert meeting on tropical forests (held in Rome in January 1982), the following is stated:

'Excessive wood removals, fuelwood and charcoal mostly for domestic use, are a factor of degradation, above all of open forest formations. Studies carried out by FAO on the fuelwood situation show, that three quarters of the population in developing countries – 2000 million people – depend on fuelwood and other traditional fuels for their daily energy needs. 100 million people are living in such scarcity situations, that they cannot obtain sufficient supplies to meet their daily energy needs: a further 1000 mio rural dwellers suffer increasing shortages and can meet their minimum needs only at the expense of exhausting available resources. The developing world as a whole suffers a deficit of 400 mio cbm of fuelwood to supply the minimum needs of people depending on this fuel.'

In the other parts of the world, in the *industrial countries* of today, a steady development led from agriculture to handicraft and various trades and, much later, to industry; and this development was accompanied by an increase in population, which found employment in the expanding branches of the economy. The consumption of energy increased enormously, partly because of the increase in population and its standard of living, mainly, however, because of the big needs of industry. This increasing demand of energy was met by the use of coal and electricity and, mainly since World War II, of oil, natural gas and atomic energy. Since the 'oil-shock' of 1973/74 a part of the rural population in these developed countries, which owns forests or lives near forests, returned to fuelwood and even some town-dwellers found the use of fuelwood advantageous again. Nevertheless, the total contribution of fuelwood to the energy balance of a developed country is rarely more than some 3 or 4%, and often less. Fuelwood consumption is limited mainly to rural households, where it will continue to be important, and to forest industries which use bark and wood residues as fuel, in so far as these residues are not of more value for other purposes (e.g. bark for horticultural mulches, sawlog slabs for fibre board manufacture).

In some countries, 'Energy Plantations' have been tried, which are somewhat similar to the former wide-spread coppice – forests. except that the rotations may be as short as 3 years or even less, but these plantations are still – regarding tree species, management, harvest, transport, storage and combustion – in an experimental stage.

The demand for fuelwood and its price have tended to increase with the result that in a few countries, notably Scandinavia, some wood of small dimensions which is suitable for the manufacture of pulp is now sold for fuel. In most developed countries, however, the rising demand for fuelwood can readily be met from wood for which there is no ready industrial market. This applies even to much of Western Europe, where many forests are underexploited and would benefit from the removal of the poorer trees in thinnings, thus giving the better trees that are left standing more room to grow. In any case, the fuelwood market is always likely to be restricted to the vicinity of forests: transport and handling costs of fuelwood are high compared with other fuels, because of its bulk and constitution. Much fuelwood is, in fact, harvested by the woodland owners themselves for their own use.

5.3.3 Policy implications

In the case of developing *countries* we must first ask ourselves what do the alarming statistics which have been quoted, signify in terms of human misery. There are first of all, the myriads of villagers engaged in collecting the fuelwood from ever greater distances as the nearer forests are exhausted and disappear. Often the wood has to be carried for long distances on foot by women and children. A few hours each day may be spent on this task alone. The food is then cooked on the primitive 'three stone hearths' which use only 5% to 8% of the wood's energy, the rest going to waste into the atmosphere. But all this is only part of the story. The over exploitation and destruction of the forests leads to soil degradation and erosion and so to poorer farm crops and pastures which, in turn, lead to over-exploitation of the soil and to further erosion. Oil is usually too expensive, if it is available at all. Plant residues and dung are used where wood is very scarce, but dung should be used to fertilize the fields. Four broad types of policy measures are needed.

1 Making better use of the fuelwood by promoting the introduction of more efficient stoves. This is mentioned first, because the benefit is immediate; the fuelwood consumption will be cut to less than half. These stoves must be inexpensive and simple, preferably made locally out of local materials and they must be adapted to local customs. It is useless to offer women who are accustomed to cook at ground level a stove where they have to cook at table

height as is customary in most developed countries. Some good designs have already been developed in various parts of the world. What is now needed is to make such stoves available where they are needed and to teach people how to use them.

2 The protection of the remaining forest areas – often open savanna or scrub – against further avoidable destruction by uncontrolled grazing, fire etc. This is only practicable with the full co-operation of the local inhabitants.

3 The establishment of fuelwood plantations within easy reach of the villages. Experience has shown that these plantations too will generally only succeed if the villagers themselves fully participate so that the plantations are *'their'* plantations. This is an important aspect of forestry in support of rural community development discussed in chapter 5.5. Much research and development work in this field is still needed especially for arid zones. The national and international research projects on finding the most suitable species and provenances and discovering the best ways of growing them should be strengthened.

4 The fuelwood question should not be considered in isolation, but in the broader context of forestry's role in rural land use and in supplying also other forest products for domestic use (wood for construction, fodder for cattle etc). Agroforestry, discussed in chapter 5.6 is one aspect of this wider problem.

Fundamental for the success of these and any other measures to improve living conditions is a good *extension* service staffed by *local* people who speak the local *language* preferably as their mother tongue and who have the villagers' confidence; they must not be seen as the 'long arm' of government bureaucracy and interference.

The urban and industrial role of fuelwood in developing countries must also not be ignored.

In towns the introduction of electricity, gas and oil as energy sources for cooking and heating abviously presents fewer difficulties than in the country where populations are more scattered and transport is more difficult and expensive. Nevertheless, where land is available, there may be a case for establishing or expanding fuelwood plantations for the production of charcoal for urban use especially in countries without fossil fuels. The Philippines have gone a step farther by establishing fuelwood plantations for the generation of electricity.

A major example of the large scale use of industrial fuelwood is provided by Brazil, where charcoal is employed for smelting steel and wood is converted to methanol as a motor fuel. Examples such as these deserve careful study by other countries short of fossil fuels.

In more *highly industrialized countries* the policy implications are somewhat different, and depend on the type of energy that is to be produced.

The demand for *domestic firewood* will continue to be influenced by the prices of alternative fuels, but for the reasons already given, the demand will be confined to the vicinity of forests. The attraction of firewood has been increased by efficient new types of wood burning stoves and of central heating systems which permit the switching between wood and other fuels. Wood based heating systems for larger buildings, such as hospitals or even for groups of buildings, have also been developed. The main policy aspects of these developments are to ensure that they receive the necessary publicity and that any additional research and development work which may be desirable is encouraged. If the luxury use of *charcoal* for barbecues continues to spread, it may be a slight boost for a few woodland owners and charcoal burners, but without policy implications.

The burning of bark and other wood residues as a fuel in *wood processing industries* has been mentioned. Further developments are largely a matter for the industries themselves and will depend on whether these residues are worth more as fuel or as a raw material for conversion into saleable products. The generation of *electricity* with fuelwood may be worth considering by some oil importing countries where land is available and the climate is favourable to tree growth. Whether or not in such cases specific 'energy' plantations are preferable to the use of the residues of conventional forestry is a debatable point. As 'energy' plantations are still in the experimental stage and as conventional forestry provides much more flexibility because of the range of the resulting produce, the advantage at present appears to lie with conventional forestry, which is also likely to provide the woodland owner with a higher income, since sawlogs fetch a much higher price than fuelwood. There is, however, a strong case for continuing and possibly strengthening the research in progress in order to see whether the very high yields that are theoretically possible with energy plantations can be translated into economically viable practice. Yields of up to 15 tonnes dry matter production per ha and year have been obtained in some experiments; in energy terms this is equivalent to about 5 tonnes of oil. A potential source of wood for fuel which has received far too little attention is the wood from old buildings that are demolished. Most of this is at present just burnt on site.

The conversion of wood into *transportable fuels*, such as methane, methanol, ethane, ethanol is *unlikely* to be introduced on any significant scale in highly industrialized countries. Much energy potential is lost in these conversions: about one third in the conversion to gas and another third to the liquid fuel. That is one reason why these products can be made far more economically from fossil fuels, the price of which would have to rise much further before that situation changes. As far as ethane and ethanol are concerned, forest trees are said to provide a less suitable raw material than some other plants. Technical developments may of course modify the situation.

The use of wood to propel motor vehicles is of course not new. Already during World War II, when other fuels were scarce, so called 'wood gas' generators mounted on a vehicle were fed with wood chips which were gasified to fuel the motor. Recently, there has been some revival of interest in this technology for farm tractors, but oil based fuels are much more convenient.

Overall, wood contributes only about 2% to 4% of the total energy consumption in developed countries and even if this could be doubled by the various possibilities indicated, the result still looks very modest. The important point is, however, that locally the impact is much greater; and the greatest impact is in remote rural areas which are usually also the poorest and where alternative sources of energy may be most expensive and not always easy to obtain.

What is, however, possibly even more important in developed countries is the role of wood as an *energy saver*. The energy input for the production of wood is sunshine only. The energy input for the processing of wood is a mere fraction of that needed in the manufacture of bricks, steel, cement and aluminium-materials which are interchangeable with wood for certain purposes. Furthermore, because of its excellent insulating properties, wood can save much energy when it is used in buildings. These energy saving properties of wood are not yet always fully appreciated by architects, builders and their clients. It should be an objective of forest policy to make these valuable properties of wood better known.

5.4 Tropical moist forest

5.4.1 Introduction

Tropical moist forest is in the focus of interest not only of the growing number of people living in or in the surroundings of these forests and the timber companies but also of wide sectors of the public elsewhere in the world. The interest has been prompted by the continuing destruction of these forests and the resulting consequences for local populations, timber supplies and the world's environment. Because of the environmental concern, the issue was included as a special topic in the World Conservation Strategy 1980, with a call for immediate action followed by a primates programme as a combined effort with the World Wildlife Fund. What are the salient facts about the tropical moist forest?

The precise area of tropical moist forests is not known, because of incomplete statistics and problems of definition. However, for our purpose, orders of magnitude will be sufficient.

Roughly 2000 million ha or 50% of all existing forest on earth is situated in the tropics.

According to Sommer (1976), tropical moist (mixed) forest (TMF) still covers 935 million ha. This area is diminishing at an annual rate, generally estimated at around 10 million ha, although individual estimates range between 5 million and 20 million ha. The approximately 900 million ha TMF which remain, account for 20% of all existing forests on earth.

The need for urgent policy action is further highlighted if TMF is considered as a biome. According to Myers (1980) this biome is reputed to cover 1600 million ha.

Consequently, only approximately 55% of the potential area exists nowadays. The definition of TMF is fundamental. Myers summarizes the biome as:

> 'evergreen or partly evergreen forests, in areas receiving not less than 100 mm of precipitation in any month for 2 out of 3 years, with mean annual temperature of 24 °C and essentially frost-free; in these forests some trees may be deciduous; the forests usually occur at altitudes below 1.300 m (though often in Amazonia up to 1.800 m and generally in South East Asia up to only 750 m); and in mature examples of these forests, there are several more or less distinctive strata.'

Within this very broad definition several forest formations may be differentiated. This is to be expected given the wide spread of TMF, (26% South East Asia, 18% Africa, mainly Congo-basin, and 56% South America, mainly Amazonia); ecological factors; and evolutionary aspects. For our purpose two salient characteristics of the TMF definitions above should be mentioned.

First, there tends to be relatively limited seasonality, notably with respect to temperature fluctuation and precipitation. Limited seasonality is the main differentiating factor between TMF (Synonym: rainforest) and seasonal (monsoonal) forest. Nevertheless, in various countries the forests with a marked dry season (seasonal forest) intergrade completely with TMF.

Secondly, biotic diversity is a main TMF characteristic. The TMF biome is, biologically and ecologically speaking, the most complex and diverse biome on earth. The complexity and diversity is dynamic and stable, though this dynamic stability can only persist within a narrow amplitude of environmental fluctuations. Indeed, recent research has indicated that TMF is likely to be more fragile and tends to be less resilient to change than relatively simple and robust temperate forest ecosystems.

With respect to the extraordinary richness in species (not only trees), Myers (1980) concludes from recent research in this field that nearly half of all plant and animal species on earth are contained in the TMF-biome although it comprises only 7 per cent of the total world land area. He adds to that, only about 300,000 of these species – no more than 15 per cent and possibly much less – have been given a scientific name, and consequently most organisms are totally unknown.

Moreover, many of the species have highly specialized ecological requirements; many exist at low densities, and many are confined to limited areas. Consequently, a forest tract does not have to be entirely cleared for a number of its resident species to become extinct. A serious reduction of life's diversity including irreversible loss of (still unknown) economic opportunity is the terrible prospect of TMF decline.

It would be beyond the scope of this book to give more detailed information on TMF, which may, however be found in the literature cited. We may conclude this introduction by recalling that a vast area of TMF still exists, but is deteriorating and even diminishing at an alarming rate, whereas its components and the potential benefits for all life on earth including mankind are still largely unexplored.

5.4.2 History

Until this century, tropical moist forest was hardly touched by mankind except by sparse populations which had lived in TMF since millenia. In those times the use by man of TMF was apparently limited to gathering produce and practicing shifting cultivation within the limits of the recuperative powers of the forest.

The development of tropical timber exploitation was limited in the main to other and more uniform tropical forest types than TMF and to manmade plantations. The heterogeneity of tree species, the difficult access and the lack of technology were constraints which resulted in vast areas of the TMF biome being left untouched by timber exploitation.

Nevertheless, from relatively early times onwards some TMF was converted into arable land and in more recent times to commercial crops such as rubber or oil palm. This shift happend according to population concentrations and to soil quality adapted to these land use forms, well to be distinguished from exploitation of the TMF as such.

After the second World War, a disaster which did not wholly bypass the TMF areas, the scene altered considerably. Some components of the changes affecting TMF deserve to be highlighted. In the course of the decolonization processes, newly established governments faced the enormous task to develop national economies. As a matter of course, the available fossil and renewable natural resources had to play their part.

Following the developments in the industrialized countries in the more temperate zones and to some extent encouraged by these, the TMF became considered as a base for economic development. It seemed inexhaustible – for the time being.

In parallel with this way of thinking, the required technology became available in the post war period. The inaccessibility of some of TMF was no longer a serious constraint and the promotion of several timber species accelerated their exploitation. One spoke even of a timber boom.

Meanwhile, in another field of development things were changing too. The progress in health care was followed by a population explosion which altered the former long lasting balance between populations and their environment, and also had repercussions for the TMF.

As already mentioned, in former times the shifting cultivation and other human interventions in TMF were able to satisfy the people's basic needs, culture and tradition without altering the forest environment. The rate of population growth, however, outstripped developments in land use practices, The result was forest deterioration and destruction on an enormous scale and at an accelerating pace. Furthermore, in some TMF regions (but not all) timber exploitation reinforced this decline of TMF areas. The encroachment was most serious in tropical Africa and South East Asia but has also become significant in the Amazon basin.

Land use planning and policy with regard to forests are dealt with elsewhere, but the question of timber exploitation in TMF must be considered here. The issue must be seperated into the exploitation of the still existing more or less virgin TMF area on the one hand, and forestry practices in other parts of the biome, mainly man-made forests, with either artificial or natural regeneration after clear-cutting, on the other hand.

As a rule, *virgin TMF is exploited* by the extraction of the trees of merchantable dimensions and belonging to commercial species, the number of which is increasing.

The typical structure and characteristics of TMF make it inevitable that the extraction will damage the remaining forest. Recent research indicates that almost any forest is sensitive to interference with one of its components (Ulrich, 1981). This statement is particularly applicable to TMF. The extraction of trees on any commercial scale represents in the TMF a far worse interference than the damage to the remaining trees, and there may be a long lasting or even irreversible change in the forest ecosystem brought about by the combination of the interference with the forest structure and the physical effect on the soil.

The principle of sustained yield can be applied in TMF by regulating first the number and size of the trees of each species that may be felled in any one cutting operation and secondly, the interval between cutting operations which is called the cutting cycle. For example, this cutting cycle in Dipterocarp forests in South East Asia is generally fixed at 35 years and the periodical cut varies from 20 to 60 m^3 per ha out of a standing volume of between 250 and 500 m^3.

In several regions silvicultural practices have been applied since the 1920's, with the aim of increasing the proportion of commercially desirable species, the Malaysian Uniform system (*Dipterocarps*) and the Nigerian Shelterwood system (*Meliaceae*) are examples. These systems include certain pre- and post-exploitation treatments, sometimes supplemented by enrichment plantings of the desirable species. Because of the complexity of the situation, the regeneration and subsequent management of exploited TMF tends to be difficult and costly. Until now the systematic management of TMF is limited to one million ha at the most.

The great variety of species which also differ in their wood properties (colour, strength, fibre length etc.) presents the wood processing industries with major technological and marketing problems. These may be reduced, but not eliminated, by grouping species for purposes of processing and marketing by their wood properties instead of their botanical affinities.

Because of the difficulties of managing TMF, *plantation forestry* has come into favour as an alternative. Some plantations are of indigenous species, but most are of fast growing exotics, such as *Pinus* spp and *Eucalyptus* spp.

An extensive example of such a plantation program exists in Amazonia, where at Jari about 100,000 ha has been planted, *Gmelina arborea* included. This type of forestry with short rotations of 15-30 years is aiming for industrial raw material rather than high quality timber logs. Plantation forestry can thus be complementary to TMF sustained yield forest management, the latter aiming for high quality timber. The present plantation area in the TMF biome is estimated to be of the order of one million ha.

However, it is too early to judge to what extent plantation forestry in the TMF biome will be sustainable in the long term especially when ecological factors are considered.

An extremely important point about the plantation alternative is that TMF is not necessarily the only land available or even the best. There are vast areas of savanna, scrub and grasses of various kinds (e.g. *Imperata cylindrica, Eupatorium pallescens)* which should be considered in the first place. Well over a million ha have already been planted.

5.4.3 Policy considerations

A sensible policy for TMF can only be developed in the context of a general forest policy which in turn must be developed in the context of a country's land use policies and general economic and social policies. It is particularly important to bear in mind that there may be much unused land outside TMF which is suitable not only for plantation forestry, but also for agriculture. Food production can often be raised significantly by making use of such land and by improving agricultural practices generally; the need to clear TMF may thus be reduced or even eliminated. It cannot be emphasized too strongly that the best way to preserve TMF is to take the pressure off it. Legislation and enforcement measures, which are also needed, can only be effective where the pressure is not too great in the first place.

It follows that the future of TMF depends on the close collaboration between the forest administration and other government departments at policy level and between foresters and experts in a veriety of disciplines at management level.

The preservation of TMF involves some financial sacrifice. Before considering how much TMF should be preserved we must seek to answer the question who should bring that sacrifice. Most TMF is situated in poor countries. Should they be expected to bear the whole burden? There are strong arguments why they should not, quite apart from the general argument that reducing the gap between rich and poor contributes to world stability and peace. First, there is the world interest in gene conservation and the world's atmosphere (see chapter 4). Secondly, some of the richest countries, which now preach conservation, until very recently 'mined' their own forest resources and may even owe some of their rapid economic development to the mobilization of that capital; these countries cannot expect others to act differently unless they are prepared to share the cost. That leads to the third reason: of the total proceeds from the logging operation only a small proportion, perhaps 5%-10% go to the country where the logging takes place and from which the logs are exported. It is the transport enterprises, the industries and the ultimate consumers of the final products in the developed countries which derive the main benefit. Finally there is the broader consideration that the conservation of TMF may be necessary to prevent erosion and hydrological disasters far from the forests themselves. Helping to repair such damage is usually much more costly than preventing it.

How can the international community effectively assist in the preservation of TMF without interfering in the national affairs of the countries concerned?

The first and most obvious way is by offering technical and financial assistance for forestry and agricultural developments outside TMF so as to reduce the pressure on it. Both bilateral aid schemes and programmes by the international organizations and credit institutions have an important part to play. The environmental organizations can promote the preservation of suitable areas of TMF and a better understanding of its significance and problems.

Developed countries can also help by encouraging the countries where TMF occurs to develop their own forest industries so that they may earn more foreign currency from a given volume of timber harvested. The latter countries could strengthen the measures they have already taken to discourage the export of logs and favour the export of products. International arrangements which stabilize timber prices at a reasonable level may help indirectly. UNCTAD and the STABEX system of the EEC may be mentioned in this context.

The welfare of the people living in and near TMF should receive very high priority. They depend on the forest not only for timber but for a variety of fibres, food and fodder. In some instances their whole lives are bound up with the forest. Should they be doomed to extinction or adaptation? Should they not be regarded as the real owners of the forest, having lived there for generations, perhaps centuries? The Indians in the Amazon region, the Dayah people in Kalimantan, the Papua on Irian Jaya and New Guinea as well as the Pygmy tribes in Central Africa are cases in point; but even where there has been much contact with the outside world TMF still provides local people with many necessities of life.

We have seen that timber-exploitation and population-pressure are the two main factors which endanger the survival of TMF. Until recently the primary aspects of TMF: site-protection, gene resource, climatic influences and so on, were mostly neglected in favour of 'timber mining' with the aim to cash in on the rich forest capital. Under favourable conditions, the secondary forest which developes after such exploitation can fulfil some of the mentioned forest functions. However, in that secondary forest the original richness, stored by long lasting undisturbed and largely unknown processes, are gone. From recent investigations it appears that, if TMF is to survive being exploited, at least two aspects have to be considered.

188

First, the basic constraint to which TMF logging must be subjected is to ensure the health and productive potential of the remaining stand, in order that it may regenerate within some decades. Equally, as in selective cutting systems in temperate regions, the allowable cut should not be estimated by measuring the present stock, but by the regeneration potential of the whole ecosystem. Because of the long life cycles of trees (up to 300 years or more) and other organisms associated with them, the natural regeneration periods are likely to be very long too and the effects of shortening this period by systematic exploitation has not yet been fully investigated. Projects in the 'Man and Biosphere' (MAB) programme of the United Nations are on the way, but results are not yet available.

Secondly, if TMF is to survive after logging, it may have to be protected against encroachment by immigrants from outside, who tend to suppress and swamp any original inhabitants.

5.4.4 Policy options

The policy options for TMF are broadly as follows:
- preservation of TMF in its natural state;
- extensive TMF exploitation designed to maintain its character;
- conversion of TMF to man-made forests;
- conversion to agroforestry or non forest use: agriculture or plantation crops such as rubber, sisal, and oil palm.

These options do not exclude one another; they are complementary, but a country cannot hope to arrive at a sensible overall policy for TMF unless the conditions under which each of these options is to be applied are carefully considered and decided. Before discussing the options individually a few further points also need to be made.

1 In addition to the wide variations within TMF which are attributable to site and climate it is also necessary to take into account the fact that every gradation exists from completely undisturbed TMF to completely degraded secondary forest and scrub derived from TMF.
2 Decisions on the future of particular areas of TMF should not be taken piecemeal as has happened so often in the past largely because agricultural and commercial interests carry so much political weight.
3 Forest administrations in some developing countries have tended to pay inadequate attention to the problems of TMF because their forest officers have received little instruction on these matters during their professional training. This should be remedied.
4 In order to arrive at sensible decisions, it is worth distinguishing between the goals (or ends), the principles that should govern the actions and, finally the

actions best suited to achieve the goals without departing from the principles.

For *the preservation of TMF in its natural state public ownership is essential.* Experience has shown that there is no effective alternative. The most suitable areas to select are those which are least accessible and least subject to population pressures. Once population pressure has started it is usually too late to prevent encroachments. In order to reduce subsequent pressure it is best to bypass such areas when major new roads are built. Even so, in the long run, pressure can only be avoided if proper arrangements are made outside the protected forests to improve farming practices and increase forest production so as to meet rising demands. Ultimately no safeguards will work if population growth continues at anything like present rates.

The size of any reserve must be based on the area required to ensure the survival of the particular ecosystem and all its component species of animals and plants. It is the species with the largest area requirements that is decisive. As these requirements are as yet not accurately known, an ample safety margin is advisable.

The extensive exploitation of TMF on a sustainable basis requires, as already mentioned, a high degree of professional skill, it is costly; it sometimes increases the danger of further human interference and encroachment; it is most likely to be successful where it has already been practiced for some time, and all concerned- foresters, government and the local inhabitants have learned to operate the system and to like it. Experience has shown that it is only in forests managed and effectively controlled by a state forest service that there is a reasonable chance of success. Where land is scarce, an important question is the relative priority which is to be given to this form of management, the main product of which if high quality timber, as opposed to the production of a larger volume of utility timber by plantation forestry. There is, however, also another consideration. Extensively managed TMF produces as by products many necessities of life for local people, fibres for basket work and rope, fruit, medicines etc.

The conversion of TMF to man-made forests is particularly suitable for forests where the original TMF has already been modified significantly by exploitation or shifting cultivation or both, but where the productive capacity of the soil has not yet suffered too much. 'Man-made' in this context may include enrichment planting or full-scale plantation forestry after clear-cutting. This system not only leads to the production of a large volume of industrial wood in a relatively short time (10-30 years) but it also creates more employment than extensive TMF exploitation, thus relieving some of the pressure on the TMF which is to be preserved: moreover plantation forestry for industrial purposes is not necessarily imcompatible with the production of fuelwood and other wood for domestic purposes as well as some fodder and food. Since the long term effects on site and susceptibility to losses from biotic and abiotic causes are still somewhat proble- mentical, too heavy a dependence on any one species should be avoided. The old adage applies 'Don't put all your eggs into one basket'.

These plantations and the wood processing industries based on them must be planned in conjunction with one another. Both require capital investment: the plantations, mainly of local currency (labour being the chief component) while the industries may need a large element of foreign currency if machinery has to be imported.

The conversion of some TMF to other land use will be unavoidable for as long as populations rise faster than the agricultural productivity of farmland. Foresters must accept this, but they should seek to make conversion subject to the following conditions:

– Each conversion should be part of a coherent land use policy which also provides for the improvement of agricultural practices on land which is already farmed; unused land outside forests which could be used for agriculture, fuelwood plantations and so on should be used before encroaching on the forest.
– Adequate provision should be made to enable the settlers on such land to grow the forest products they need for their own use; agroforestry, shelterbelts, fuelwood plantations etc. all can play a part.
– The maintenance of long term soil fertility must be assured.
– The timber on land to be cleared should be fully utilized instead of being simply burned and wasted as so often happens. A well planned 'timber mining' operation is called for; this can be commerically attractive too, provided that care is taken that the proceeds go into the right pockets.

In conclusion, the policy decisions on the future of TMF should be taken within the framework of a general land use policy, bearing in mind that the best way to protect TMF is to promote good forestry and good agriculture outside the TMF. Environmental, social and economic factors must be duly weighed but with emphasis on the long term ecological considerations. The interests of the local people should receive priority over distant urban interests. Above all, the time for action is now while there are still substantial areas of Tropical Moist Forest to be saved.

5.5 Farm forestry

5.5.1 Introduction

Farm forests have been of great benefit to farmers and have contributed significantly to timber production in some developed countries and they could prove equally beneficial in others where this category of ownership is rare or non-existent. They may also become relevant in some developing countries when appropriate systems of land tenure and agriculture have become established. It is for these reasons that a separate section is devoted to farm forestry.

Farm forests play an important role in countries with both a strong farming and forestry tradition, such as is found in Central, Northern and in parts of Western Europe as well as in parts of the United Stated of America. In some countries with centrally planned economies such as Yugoslavia and Poland, farm forestry has also considerable weight. On the other hand, in Great Britain and in Ireland farm forestry is rare, because in the past most farmers were tenants of big land-owners, who leased the farms, but retained the management of the woodlands. When these large estates were split up, the farms were usually sold separately from the woodlands.

Even in countries where most farmers own forests and a large proportion of the total forest area is in this category of ownership, exact figures rarely exist because the statistics do not differentiate between farm forests and other privately owned forests. This is unfortunate because the distinction may have policy implications.

Farm forests are often called the 'second leg' of a farm, but as far as for instance big mountain farms are concerned, whose forest area can be 200 or even 500 hectares, forestry is often the first and most important 'leg', which contributes decisively to the subsistence of the farm. The farm forest can, of course, be of little importance to the farm, if its size and its share in the total area of the farm is small so that perhaps only the farm's own consumption of wood can be covered.

But often it is not wood production alone. In order to obtain a satisfactory family income, the possibility of making a better or even full use of all the labour force of the farm, be it human or animal or mechanical, may be equally important, and this is facilitated because the work in the woodlands can usually be done outside the peak periods of work in agriculture; the total work-load is spread over the seasons and no temporary umemployment occurs.

Thus, a big and well managed farm forest can be the most important economic branch of the farm, important by its production of wood for sale, important also of course with regard to the supply of wood for home consumption, important however, furthermore, as a source of labour income, when the farmer with his sons or with the help of neigbours carries out all forest work himself and earns, thus, the high wages of a professional forest worker. In many countries if there is a surplus of labour capacity on the farm, the men work for a certain period in the year in a neighbouring forest estate. In Scandinavia about half the timber harvest is said to be carried out by farmers.

Last, but not least the fact must not be overlooked that it was the farming population which has built and formed in a thousand years the so-called cultural landscape with its mixture of fields and woods, which is beautiful and has a great recreational value.

Very much depends, of course, on the weight a country places in its general policy on private property, on independent farms and family enterprises and on the esthetic value of the landscape.

5.5.2 Types of farm forests

Three different types of farm forests should be distinguished:

- The *'farm forest proper'* (bäuerlicher Eigenwald) which belongs as a privately owned property in the fullest sense to the farm. The farmer, and often his wife, are the only and full proprietors and they can manage their forest as they like within the limits of the law. Fellings and timber sales can be adjusted much better to market conditions than on a large pure forest estate, because within the farm the forest is one branch only; the farm and the owners do not depend on the forest alone.
- The *share in a 'common forest'* (Gemeinschaftswald), which belongs to a number of farms. The forest has to play here only a supplementary role. The management of these common forests depends mainly on existing regulations and on decisions of the majority of all part-owners. The influence of the individual farmer and his 'freedom of action' is therefore very limited. As a rule he has to take his share of wood annually; 'savings' in the form of standing trees (as a higher growing stock), such as he could make in his own forest, are, thus, not possible.
- *'Rights of usage'* (Nutzungsrechte in fremdem Wald) for wood, grazing or forest litter (Waldstreu) in a big, privately or publicly owned forest. Here also the farmer who owns this right, gets his wood on the stump every year and has to fell and to remove it in time. No savings are possible, which often would come in very useful in cases of misfortune like fire, illness etc. or when costly repairs or investments have to be carried out or when the farm is given over to the next generation and the other children have to be paid out. In all these cases the forest is used as a 'savings account', but only the 'farm forest proper' can act in this way.

It would be quite wrong of forest policy when dealing with the farm forestry problem to take into account only the 'farm forests proper'. This would mean to see the farms not in their full extent. It is only together with the two other types of farm forestry that farms appear in their right size and viability; these other two types may contribute up to 30% in forest area in some localities.

As a matter of fact these three types of farm forests can be found in all combinations, either as 'farm forest proper' alone or combined with one or both of the other two or as 'share in a common forest' or as 'right of usage' alone or combined with the others. In Austria for instance, the area of farm forest proper of about 1.75 m. ha is supplemented by 0.33 m. ha of common forests and by the equivalent of the rights of usage corresponding to 0.20 m. ha. This means in some Austrian provinces an additional forest area of 8 to 12 ha for every farm.

There are, of course, also farms without a forest of any kind. These farms are of special interest insofar as their home-consumption of wood is much lower, often one third only of that of farms with forests.

5.5.3 The dynamics in farm forestry

There has been a gradual *change in the multiple use* of these forests. In former times, but not very long ago, forest grazing and the collection of forest litter as a bedding material for domestic animals have often been as important as timber production. In many countries these secondary uses have almost or even completely disappeared. They have, however been replaced by other secondary uses which benefit the public like protection, landscape conservation, recreation etc. The consequences of the former secondary uses, which have been practiced for generations, have however by no means been overcome. The trampling effect of forest grazing, the impoverishment of the soil by the collection of forest litter, the subsequent change in tree species composition etc. etc. make themselves felt on hundred thousands of hectares. It is an important task of forest policy to get these problems investigated and researched and to provide for a good extension service and possibly for subsidies too.

Farm forestry may offer a partial solution to some of the problems brought about by modern farming practices. Much former manual work on farms is now done by machine. In consequence, former agricultural land which is unsuited for machines because it is too steep or too stony has been afforested by farmers. But modern farming methods have so greatly increased farm production, that even where constraints do not apply, some land, usually the poorer land, has become submarginal for farming but may still be very suitable for forestry. The displacement of manpower by machines on the farm has not led to a surplus of manpower because more and more young people left farming for jobs in factories. In fact in some rural areas there has been a growing shortage of manpower for agricultural and forestry work.

The exodus was accelerated by the fact that small farms could no longer provide full time work and an acceptable standard of living for a farmer and his family. Many therefore sold out; farm amalgamations were in fact encouraged and subsidized by Governments.

In countries with a tradition of farm forestry farmers have tended to afforest themselves the land no longer required for farming or to sell it to other farmers with an interest in forestry. Where there was no such tradition as in Britain or in Ireland it was the state and private absentee investors who were in the market to by such land; farmers showed little interest and were not encouraged to do so by the government; this was probably a mistake because it led to a divorce of agriculture and forestry, when a marriage could produce so many mutual benefits. An argument that has been adduced against the encouragement of farm forestry has been that it merely leads to inefficient small scale working by people without much forestry experience. The essentially narrow economic advantages of this approach are more readily visible than the broader advantages of closely integrated rural land use and of farms standing on 'two legs' instead of only one. It is also worth noting that there is now considerable evidence that dedicating part of the area of a farm to forestry does not entail a corresponding reduction of agricultural output; there may even be a slight increase.

The *shift from full-time to part-time farming* makes itself very much felt in some countries and influences farm forestry often to a high degree. This development has been going on mainly in highly industrialized countries especially since the 1960's and has not come to a stop yet. In some countries the portion of part-time farms has risen to more than 50%. The reason is that farming is in many cases, especially under unfavourable conditions as for instance insufficient area, poor soils, rough climate, undeveloped infrastructure, etc, not profitable enough, to allow the farmer and his family to keep their standard of living in step with the rest of the population. Thus, they try to improve and increase their income by part-time farming, that is, they work at the same time in another profession, too. If this is done on a limited scale only, it may be very favourable and these farmers are at the same time a valuable link between the agricultural and the industrial population. However, when the other occupation is taking on the character of the main profession, this may then often mean an absence of the farmer from his farm, either daily, except perhaps the weekend, or weekly or even longer. The result is then on the one hand a considerably higher income, compared to which the income from the forest may lose much of its former weight; on the other hand the farmer can help to run the farm only in the evenings or at weekends, the main burden of work being left to the farmer's wife who now has not only to look after the household and the children, but also after the cattle and fields and the forest too. The consequence can only be; an overstressed farmer's wife and some loss of efficiency on both the farming and the forest side of the holding. One solution may be to switch some of the agricultural land to forestry which requires less labour and permits more flexibility in the timing of work.

5.5.4 Forest policy and farm forestry

Some of the policy options have already been mentioned. The basic question is to what extent, if at all, a government should take active steps to promote farm forestry. Where it already exists there is an obvious case for securing its future. Where it does not exist, the case is less obvious but certainly worth examining carefully and objectively; this is not easy for two reasons. The one has already been mentioned, namely that the benefits are less easy to assess than the disadvantages. The other is that neither forest authorities nor agricultural authorities tend to show much enthusiasm for what is essentially a joint venture which might make life a little more complicated for each.

In all these industrialized countries, the handicap of farming is, that agriculture is producing for an only slowly or even not growing market. If the farmers try to increase their income by producing more, the result is always an overproduction with all its difficult consequences. This danger of overproduction does not exist with regard to forestry and wood. For agriculture the growing of farm crops for the production of energy (bio-energy) may perhaps be a partial way out of these difficulties in future but for the time being, forestry is the obvious answer.

Apart from other considerations, if the farmer himself does the afforestation, he does not have to relinquish the land and land ownership means still a great deal to many farmers, especially if the land has been owned by the same family for generations. Such a deep-rooted attachment to the land should not be ignored lightly where it exists, because it promotes political stability.

We must now consider the specific policy measures that should be considered for the promotion of farm forestry. They are in the main:

- *Dissemination of forestry knowledge*, in order to get the necessary measures done in the right way, and thus to secure a good success, which is always the best stimulus to carry on.
- *Financial support* to facilitate these measures and to give at the same time a certain reward for the unpriced, but very valuable services rendered to the public by farm forests. In parts of Western Europe and especially in the EEC the financial support needed for farm forestry would be small compared with the support which is actually given for agriculture and, in some instances, leads to overproduction. Support for farm forestry could thus lead to a net financial saving.

Where farm forests represent a big part of a country's forest area as a rule much can be done to increase production and productivity of these forests; this should be a strong incentive for those responsible for forestry policy to deal intensively with this category of forests. However, this should not be the only reason to regard farm forestry as worthy of support; another reason is, that farm forestry is connected with farming and thus with a part of the population producing much more than only agricultural and forest produce. In the course of centuries this farming population has produced the present landscape, which we like and love, in which we feel at home and which is often of high attraction to foreign tourists and, thus, the basis of an important trade. Our concern should, therefore, not be the forest alone, but also the farms and the people living and working on them, and the 'cultural landscape', which should be developed smoothly and not changed abruptly.

Since it is the farmers who will have to implement any policy on farm forestry, they must be encouraged to play their part in policy formation; *their* interest has to be awakened, *they* need to be informed and to be helped.

In many cases co-operation among the owners of farm forests can be very helpful, first of all in forest road building, where co-operation is almost indispensable, then in marketing, where co-operation can be very helpful to overcome the weak position of a small supplier to the market.

A delicate question is how far forest policy should go in promoting co-operation among forest owning farmers. Theoretically, if many small forest holdings are managed as a single unit, this should result in more efficient management but experience has shown that this is not necessarily so and that the old saying may apply 'quod communiter geritur, communiter neglegitur' (what is managed in common, is neglected in common). Much depends on the people involved. Where there is no strong and sensible chairman, supported by an efficient executive committee, things can easily get worse instead of better. The disadvantages for the individual owner of losing some of his freedom of action must also not be overlooked. Finding the right intermediate policy between perpetuating extreme individualism on the one hand and promoting exaggerated and compulsory management in common on the other hand is not easy. It will nearly always pay not to move too far too fast, to be flexible in approach and to concentrate in the first place on a few key issues. One such issue is to ensure a sound information base: good forest inventories, market information etc. Another of importance in some areas are schemes to promote land consolidation. Where farmers tend to have several scattered small parcels of forest (usually resulting from certain laws of inheritance) it would be to their advantage and it would facilitate forest management if by exchanges of land, this fragmentation could be reduced. The problem of course is not easy because each patch of forest differs from every other in soil fertility, growing stock, access etc.

To sum up, the objectives of policy with regard to farm forestry should be to establish the relevant facts, to stimulate interest, to disseminate knowledge and, if necessary, offer guidance and financial support.

Indispensable prerequisites are

> a good *Forest Law* – an efficient *Forest Authority* – a very good *extension service*, which may often be joined with agriculture and which in addition to offering advice, has to carry out all the necessary promotion activities like courses, demonstrations, excursions, competitions etc.

5.6 Agroforestry

5.6.1 Introduction

Interest in agroforestry has increased enormously in recent years. A major new organization, the International Council for Research in Agroforestry (ICRAF) has been established in Nairobi (Kenya) with the aim of developing and spreading the application of this type of land use. A great deal of attention has also been devoted to the subject in technical journals and at international conferences.

Agroforestry has been *defined* in various ways. In this book the term is used in the broad sense of any form of land use which combines components of forestry, agriculture and/or animal husbandry on one piece of land or within a small farm. It is thus a new name for a wide range of old and new systems of land use and encompasses everything from forest grazing and the intercropping of trees with food and fodder crops to the growing of trees as shelter or for timber in fields and around homesteads. The methods of integration depend on the site as well as on local needs and customs.

The main *objective* of agroforestry is to enable farmers to secure the maximum *sustainable* yield of the agricultural and forest products which they need for their own consumption and for sale. The emphasis on *sustainable* signifies the importance that is attached in modern agroforestry to maintaining the long term productive capacity of the site.

The main *need* for agroforestry arises where forests and land for agriculture are scarce and no suitable land is available to create new forests; but even where there is no shortage of land farmers, may derive, under certain circumstances, more benefit from *modern* agroforestry than by keeping farming and forestry separate. This applies especially to developing countries, but also in developed countries agroforestry is regaining some of its previous importance. This is because while some of the older forms of agroforestry were very harmful to forestry without bringing corresponding advantages on the farming side, modern agroforestry can be beneficial to both forestry and farming.

Within the broad concept of agroforestry various specific systems have been given names such as (agro)silvipasture, agri-silviculture, silviagriculture; there are also traditional local names for particular agroforestry practices such as taungya and pekarangan. Sometimes different names are applied to a particular practice or, conversely, the same name is applied to different systems. These and the many other technicalities involved do not concern us in the present policy context.

From the point of view of trying to improve the living conditions of hundreds of millions of poor people the following policy aspects of agroforestry are of relevance:

- optimum sustained yields to be obtained by the improved edaphic and micro-climatic conditions under simulated forest conditions;
- the integrated production of agricultural and/or animal products, tree crops and other (forest) plants or services;
- management techniques compatible with local customs and traditions.

While there is clearly a strong case for the encouragement of agroforestry, this must not be at the expense of other aspects of forest policy. Agroforestry can be an extremely valuable tool of forest policy but it is not an answer to all forestry problems. Exaggerated enthusiasm could do as much harm as neglect.

5.6.2 History

Agroforestry has a long history in both developed and developing countries.

Shifting cultivation, the slash and burn agriculture which is still common in many tropical regions was in former times also practiced in other parts of the world. There is some evidence that already the ancient Celts may have practiced similar systems in Europe in some parts of which these practices persisted well into this century, e.g. the 'Brandwirtschaftkultur' in Austria.

But also more sophisticated systems of agroforestry are believed to have been developed a long time ago, for example by the Mayas in Central America and the Incas in the Andean region of South America. Some of these better systems are still in use today, in various parts of the world. In the mountainous areas of Java, for example, the wet rice fields and dwellings of each household in the valleys are managed in close conjunction with forest gardens higher up the slopes, which contain fruit trees and various dry land forest crops. Translated into modern terms, appropriate land-use combinations of woodland and agri-pastoral components are energy circles: energy from living systems which absorb the sunlight, is tranported to other components, partly recycled but partly stored in arable land by means of dung and plant residues.

It is worth noting that many features of European landscapes which are nowadays mostly appreciated for reasons of aesthetics, ecology or nature conservation, are in fact elements of former agroforestry. Hedgerows, wooded banks, small copses and individual trees in pastures are examples.

Another historic form of agroforestry in parts of Europe was the practice of intercropping in which rows of tree seeds were sown between rows of foor crops. The main purpose was afforestation; the trees benefitted from the tillage and weed-control which was required for raising the food crops. The system was sometimes also intended to reconcile peasants to the idea of reafforestation where forests had been overexploited or destroyed. The so-called 'taungya' system which has already been mentioned, bears some resemblance. This system originated in Burma and was introduced to other parts of South East Asia over a century ago and is still practiced, for example in Indonesia (tumpang sari).

Agroforestry thus clearly presents no new concept, but it has recently acquired a new significance after many decades of decline. What brought about first the decline of agroforestry and then its subsequent move back into popularity?

The reasons are partly technical and partly human. In *developed* countries progress in agriculture involved the use of ever more and larger machines; trees are an obstacle to these machines. Better pastures could be provided outside the forest than inside it and without risk of damage to the forest. The more intensive management of forests too was made easier by concentrating on the trees alone; the practice and teaching of silviculture and of agriculture evolved accordingly. That leads to the human dimension. Both forestry and agriculture became more specialized; foresters and farmers no longer pursued common interests and they went their separate ways.

At farm level it was easier for the farmer to increase the production from farming and from forestry if he kept them separate. Where a conflict of interest arose, it was usually the forest that suffered, partly because the short term problems of food production took precedence and partly because most farmers who owned woodlands knew less about forest management than about farming. While the exclusion of grazing and food production in forests was almost entirely beneficial, the elimination of tree growing outside the forest was not. The uprooting of hedge row trees and copses facilitated mechanical farming but led to erosion, mainly be wind on flat land and by water on slopes; furthermore, the removal of these trees and accompanying vegetation deprived many species of birds and other animals of their habitats and impoverished the landscape. Even where there were no trees originally, as on the former prairieland in the USA, modern farming has led to wind erosion and the formation of dust bowls. There is thus a strong environmental interest in the planting and maintenance of trees as shelterbelts on farmland outside the forest; if these trees in addition to their environmental function can produce useful timber or other produce, so much the better. The recognition of the environmental advantages of some form of agroforestry has led to a new interest in it.

In *developing countries* the reasons for the decline and resurgence of interest in agroforestry were somewhat different. Slash and burn farming in forests did little harm as long as the period of forest fallow after cultivation was long enough, say 30-40 years, but as populations increased the period of forest fallow became shorter and shorter, so that neither the tree growth nor the soil could recover. The inevitable result: forest degradation and erosion. Even where the population pressure was less, the forests tended to become degraded except where a conscious effort was made to plant trees during the period of growing food crops as in the taungya system, which however, only worked when it was properly organized and supervised and when the shifting cultivators were given appropriate incentives not only to plant the trees, but to tend them afterwards for a few years until they needed no further protection. Where these preconditions did not exist, a separation of forestry and agriculture seemed the best solution, especially as permanent agriculture is the key to permanent settlements which in turn are the foundation for developing social services: schools, medical facilities etc. Unfortunately, what was sometimes not sufficiently appreciated when permanent agriculture replaced shifting cultivation, is that technologies which had proved their worth in other climates and social conditions might not be applicable.

In many developing countries, the case for agroforestry is incontestable. There is just no other way of meeting peoples' needs for food, fodder and wood for fuel and other domestic use, when no significant areas of land can be spared for pure forestry. Parts of Bangladesh are an extreme example of this situation. The fact that the high output of agroforestry requires a high labour input is an advantage where there is rural underployment as is so often the case in underdeveloped countries.

5.6.3 Policy aspects

Some policy implications of agroforestry are universal, others apply mainly to either developed or to developing countries.

The first question any government anywhere must ask itself is whether a policy for agroforestry is necessary or desirable. There are likely to be few countries where, in the light of what has been said above, the answer will be *no,* but the emphasis to be given will obviously vary. To regard agroforestry as a panacea would be as bad a mistake as to ignore it.

The next question is who should be the main beneficiaries of agroforestry. Above all they must be all the people who own the land and all who work on it, but they themselves must achieve these benefits mainly by their own efforts, supplemented by such assistance as may be justified by the national considerations. A thriving farming community is clearly in the national interest, and so is any contribution to a country's timber production, soil conservation and the landscape that agroforestry can make. The next question arises from the fact that agroforestry necessarily involves an interdisciplinary approach from policy level right down to its application to a particular piece of land. Unfortunately the traditional training of foresters and of agriculturists does not place sufficient emphasis on such an interdisciplinary approach; that explains to some extent the rivalries that tend to exist between the two professions. An interdisciplinary approach to training should therefore be promoted. The benefits from this will only be reaped after some years. An immediate step, however, is to decide who should be responsible for agroforestry. Should it be the responsibility of the forest authority, of some separate authority, or should the responsibility be divided in some way? A separate authority is only justified if agroforestry is to play a very important role, because any new organization complicates and adds to the expense of administration, but it facilitates the inter-disciplinary approach especially if the organization is staffed by professionals from various disciplines. Where the forest authority is made responsible, some organizational separation between those responsible for what happens inside forests and what happens outside may be desirable. Many foresters feel so much more at home inside forests than outside them, they understandably prefer to manage forests rather than help others to grow trees outside: similar considerations in reverse apply if the agricultural authority is made responsible. The recruitment of agriculturists into the forest administration and of foresters into the authority for agriculture may be worth considering. Leaving the responsibility devided may work where those concerned are able to co-operate. As with so many problems, the best solution depends on local circumstances.

In *developed countries* the specific policy objectives should be the following:

1 To discourage the continued destruction of hedgerows, clumps of trees etc on farmland, where such destruction is environmentally undesirable;

2 To promote good agroforestry practices which already exist, or where they would be beneficial. The planting of poplar trees on agricultural land is an example in some countries such as the Netherlands, France and Italy; hedgerows as shelter against wind, trees to provide shelter and shade for cattle are other examples;

3 To promote promising new practices: for example the growing of some high quality timber trees (walnut, ash, cherry) could be a pleasant hobby, bring in some money for the farmer and secure future supplies of high grade wood when virgin supplies from the tropics start to dry up.

The main policy tools are likely to be advice, guidance and demonstrations by the advisory services of the appropriate government authority and, in some instances, of the farming organizations themselves. To quote an example: it is not sufficiently appreciated that clumps of trees at corners of fields interfere very little with the use of farm machines which cannot turn at a sharp right angle; there is thus always some 'dead' ground at the corners. There may also be a case for giving financial incentives and for sponsoring research and development work on agroforestry.

In *developing countries* the policy problems associated with agroforestry are generally much more urgent and more difficult: more urgent, because of population pressures and the impossibility of meeting the needs of rural populations by other means; and more difficult because less is known of what can and should be done; effective extension services do not exist, and there is a shortage of trained personnel and money. There is also still a reluctance on the part of forest officers and agricultural officers in some countries to devote much effort to this difficult and often unaccustomed task.

A first useful step is to ascertain the relevant facts: what are the farming practices, what is the supply and demand position concerning fuel, social customs, wood for domestic and agricultural purposes, wood for sale, trees for fodder and fruit etc. Such an inventory, if well conducted, will not only provide basic statistics but will also pinpoint difficulties and problems and may even suggest some possible lines of action. Bangladesh is a country which has demonstrated that such an inventory is both practicable and useful.

The next crucial step is to provide an effective extension service. To do so takes time; any attempt to move too fast is likely to jeopardize efficiency and an ineffective service is worse than none. A suitable approach may be to concentrate efforts in the first place in districts where success is most likely to be assured, even if these districts are not necessarily the ones in greatest need. These centres can then serve as demonstrations, and the experience gained will help in dealing with more difficult areas. If a scheme is started in the most difficult areas the risk of failure is increased, a failure which may endanger a whole scheme. A programme must be carefully planned and organized and the extension workers on the ground must be people with local experience, preferably coming from the communities whom they are expected to serve. They must be familiar with both the tree crops and agricultural crops. Only so can they gain the confidence of the villagers. An extension worker must be able to advise on all the farmer's problems.

A very delicate policy problem to be taken into account is the social structure of the rural communities concerned. Apart from any other divisions, there are likely to be divisions between relatively large farmers, very small farmers and those owning no land who are generally the poorest. Any improved method of land use including agroforestry will tend to bring the biggest benefits to the largest farmers and least, if any, benefits to the landless poor, except indirectly insofar as they may have better opportunities for gainful employment and participate in a general improvement in the standard of living. On the other hand, it is likely to be the larger farmers with say 30-50 ha (large estate owners are not considered as belonging to rural communities in the present context) who are in the best position to promote the introduction of new methods. They are generally better educated than the others and exercise a greater influence in community affairs. For the sake of social justice and for reasons of practical politics any accentuation of the gap between the 'haves' and the 'have-nots' should be avoided and this may not be easy. In some instances there may still be land available for planned settlement; state land or land in common ownership of a community; possibly even forest reserves may have to be opened for the purpose; where no land is available other solutions must be found, e.g. by allowing the landless to grow trees and graze cattle along roads, dams etc. or by devising some partnership scheme with the larger farmers. No general recommendations are possible. Here it must suffice to draw attention to the problem. In some situations, the encouragement of tree planting by religious or educational institutions may help to get the idea accepted by villagers.

The question of research and development work must also not be neglected. Much clearly needs to be done but two common pitfalls should be avoided. The first is to defer action on agroforestry until more research results are available. This is wrong partly because what is already known far exceeds what is applied in practice and partly because in a situation of dire need any immediate action that prevents things from getting worse and possibly brings about a modest improvement in living conditions is better than no action at all. The second pitfall is to duplicate unnecessarily research which has already been done elsewhere. Much research on agroforestry is now underway in various parts of the world; research organizations and journals are dedicated to the subject. Clearly, research results may not apply under conditions different from those where they were obtained, but much time and effort can be saved if what has been done elsewhere is taken into account before embarking on local research and development programmes. In several countries exotic species have already proved a great success in agroforestry.

Finally, because of agroforestry's great importance to developing countries, various international organizations and credit institutions including FAO, ICRAF, UNDP, the World Bank and the Asian Development Bank as well as bilateral aid agencies have been active in this field. Developing countries about the embark on agroforestry programmes would be well advised first to contact these sources of aid.

The various actions required to introduce and develop an agroforestry programme must be carefully planned and co-ordinated if there is to be a reasonable chance of success. Subsequent progress also depends on the monitoring of what has been achieved. The danger of getting trees planted and then letting them die through neglect is particularly great when the planting is done by many people and scattered over a wide area as is usually the case in agroforestry projects.

5.7 Rural community development

5.7.1 Introduction

Forestry can contribute to the well being of the people living in rural areas in many ways. Conversely these people can play a major part in either making the forests around them more productive or gradually destroying them. Various aspects of these relationships have been discussed earlier in this chapter under forest protection, wood for energy, tropical moist forest, agroforestry and farm forests. The present section places these aspects of forestry into the broader context of rural community development.

Some of the benefits to be derived by rural communities from forestry are similar in both developed and developing countries: for example forest products for domestic use; employment and income based on the forest and on the wood processing industries, as well as on tourism. The policy implications are, however, somewhat different; the two sets of countries will therefore be considered separately, taking the developing countries first where the problems are both more difficult and more urgent because the vast majority of people depend directly on the land for their living and on wood and other forest produce for many other basic necessities of life.

5.7.2 Developing countries

The rapid increases in populations in most developing countries over the past generation have not been accompanied by a corresponding improvement in land management. As a consequence, more and more forest has to be cleared to grow food, cultivated land loses its fertility and is eroded; and the rural people have become poorer than ever. Rural community development programmes can reverse this unhappy chain of events but only if there is a co-ordinated approach to all aspects of the problem including the very sensitive aspect of persuading communities to slow down and eventually stop further population growth. A precondition for good land management, which is the only way to make land more productive on a sustainable basis, is an equitable and stable system of *land tenure*. In countries where most of the land is owned by a smal number of large, absentee landlords while the great majority of people who work and live on the land have no stake in it, little progress can generally be made without *land reform*. The question of land reform goes far beyond forest policy, but forestry and especially forestry in support of rural community development is profundly affected by the system of land tenure. Clearly, people have no interest in planting trees unless they have some assurance of reaping some benefit from the harvest. This is not the place to go into the social and political aspects of land tenure, which arouse such strong emotions but a few points of particular importance in the context of forest policy must be underlined.

Large units of land ownership, whether public or private, facilitate efficient forest management, which results in high yields and a high level of gainful rural employment; but forest management will only be efficient if the owner has the interest, the knowledge, and access to the capital required for development. Large units are particularly useful for the development of forest industries which require a large and steady supply of wood and raw material. This large scale industrial forestry contributes to rural community development mainly through the creation of gainful employment and the social services – schools, medical facilities etc. – which accompany the establishment of modern wood processing industries.

Large units in public ownership and control are also essential, as already stated in 5.4, for those areas of tropical moist forest which are either to be preserved in their natural state or which are to be managed so as to preserve a rich species mixture. Public ownership does not preclude *limited* rights by local populations to collect firewood and various kinds of minor forest produce – fruit, fungi, fibres, firewood etc.

Small units, owned, or at least managed, by individual peasant families or small local communities are generally best for supplying domestic needs of wood and other forest produce. Fuelwood plantations, agroforestry, farm wood lots all can play a part. Farms may also be well suited for growing timber for local arts and crafts and purposes such as boat building. If in addition, a surplus for sale to wood processing industries can be grown, so much the better because this generates cash income; but it is only in exceptional cases that a major industry could be based entirely on this source of supply.

A point generally overlooked by professional foresters, is the fact that many tree species can produce far more products outside than inside the forest. There is a case for the greater use of nitrogen fixing trees, fruit and nut trees for cash crops, shade trees with possibilities for fuelwood production by coppicing and pruning, trees for wind breaks, and so on. Some tree species can produce 15-30 m^3 and more per ha per year. On Java, *Sesbania grandiflora* can produce 25 m^3, *Calliandra callothyrsus* in coppice even 50 m^3 per ha per year. On the Phillipines, coppice forests of *Glericidia sepium* and *Leucena leucocephala* (ipil-ipil) with respective yields of 23-40 m^3 and 28-35 m^3 per ha per year are not exceptional. In such circumstances not only the stem wood but also the branches 2 cm and up are utilized. This type of forestry can of course also help to prevent erosion. Experience has shown that to be successful, forestry measures must be closely co-ordinated with all other aspects of rural development and that also within forestry there must be a comprehensive approach. If there is a shortage of fuel and of fodder a forestry programme that remedies both shortages and preferably also introduces fuel saving stoves is more likely to arouse interest and succeed than a programme dealing only with one of these problems. Similarly, it has been found that the best way of interesting people in growing timber trees that may take 10 or 15 years to mature is to combine such a venture with the growing of fruit or fodder-trees that take only half as long and with fuelwood species that can be harvested even sooner; if annual agricultural crops can also be incorporated as in agroforestry so much the better. In China in particular this combination of crops requiring different lengths of time to mature is said to have proved to be very successful.

Forestry can only play its part in rural development if there is an effective extension service and if there are incentives. Ideally there should be a single extension service to cover all aspects or rural development, including a strong forestry component. Failing that, there must at least be good collaboration between forestry extension and agricultural extension. The villagers do not wish to be confronted with conflicting advice. As already mentioned, also the forestry aspects must not be dealt with piece meal.

Success of forestry extension work depends mainly on three factors:

1 *Understanding the problem;* The forest workers responsible must not only have the necessary professional expertise, but they must also know and understand the villagers, their problems, customs and needs; These needs may include peripheral activities such as bee keeping, basket weaving etc. Racial affinity and a common language with the villagers are a great advantage, at any rate for the junior forestry staff. It is only after steps have been taken to meet these preconditions that effective work can begin. Forest officers accustomed to manage state forests often have little interest in and understanding for extension work.

2 *Collaboration between all concerned*: There must be very close collaboration between the forest service personnel and the villagers; this may not be easy to establish, since it depends on the interest of the villagers which may be difficult to arouse in regions without a forestry tradition; pilot projects to serve as demonstrations may help; although firm guidance by the forest service may be

needed, the objective must be a transfer of responsibility to the villagers as quickly as may be practicable, but subject to an adequate continuing monitoring of progress.

3 *Perseverence*: Many projects which get off to a good start are subsequently abandoned when the initial enthusiasm wanes and difficulties arise. Changes in personnel, both in the forest service and in the village leadership must be handled with tact and understanding if momentum is not to be lost. Projects must be allowed to evolve in step with social, economic and technological progress in the communities. Above all there must be stable and continuing government interest and support.

On the question of incentives, the important point to remember is that the time horizon of most peasants is very short; they are accustomed to sow and reap within a few months. To encourage tree planting, it therefore rarely suffices to provide the planting stock and to pay something for the labour of planting. Some further incentives may be needed for the subsequent tending. Under some schemes part of the money thus spent is recouped at the time of harvest, but normally this would either be impracticable or involve disproportionate administrative expense for collection.

Successful forestry programmes in support of rural community development have been undertaken in a number of countries, in some cases with the assistance of international agencies and credit institutions. The countries include China, Indonesia, Philippines, Senegal ('Maisons Familiales'), South Korea and Upper Volta ('Bois de Villages'). Countries which have not yet embarked on such programmes can learn many useful lessons from those who have.

In extreme instances, community development implies a fundamental change in a way of life that has remained almost constant for centuries, in fact a transition from the stone age to the 20th century. The problem is essentially a human one and if governments really have the welfare of these people at heart, their views, beliefs and even what may be regarded as prejudices must be respected.

There are still small, scattered tribes living in virgin forests, e.g. in Amazonia, Kalimantan and Papua New Guinea whose lives have as yet been little affected by contact with modern civilization; but contact is unavoidable; in the past such contacts have mostly been disastrous for such indigenous peoples. The treatment they have received from modern man has at best been insensitive and at worst verging on genocide. What should be done for those indigenous tribes that are still left? Nearly everywhere the population pressures building up around them would render any policy unrealistic which aims at maintaining the *status quo* even if that were considered to be in their best interest. What can be done is to ease the transition in various ways e.g. by allowing as much time as may be practicable, by preventing the unscrupulous exploitation of such people and by carefully conceived programmes of education and improved methods of land use. The leaders of these communities should be given every encouragement to bring about necessary changes themselves with the minimum of outside interference and with the retention as far as possible of tribal customs, culture and language. Well meaning paternalism is often resented and may be counter-productive. In some parts of the world indigenous minorities have become a tourist attraction. Tourism may be beneficial insofar as it may give an incentive to retain cultural traditions of dress, dance, song, artesania etc. while at the same time promoting contacts with the outside world. On the other hand ill mannered tourists may breed resentment; this cannot be in the best interests of rural community development.

The much more frequent situation is where rural communities have already been in touch with the outside world for generations but have continued to a greater or lesser degree with their old systems of shifting cultivation which because of rising populations is ruining the forest and eroding the land. There is here no alternative to the rapid improvement of agricultural practices but this must be done in a way that the farmers find acceptable. To act otherwise is neither civilized nor, generally, effective.

5.7.3 Developed countries

In most developed countries there was until very recently a steady migration of the population from the land to the towns which became even larger and more crowded while some rural areas became more and more depopulated. The main reasons for the migration were the hope of better living conditions: better paid jobs in industries; social progress, more free time and entertainment, better educational facilities for children, proximity to shops etc. For many, especially the poor, these hopes were not fulfilled. They had over-estimated the 'quality of life' in towns. As a result, there has been a move to the suburbs and a decay of town centres. Many of those who can afford it have acquired holiday homes out in the country.

How does this affect forest policy? The exodus from the countryside has led to a shortage of forest workers in some areas, although mechanization has greatly reduced manpower requirements in forestry as it has done in farming. Most people nowadays who prefer life in the country to life in the city, no longer want to live in isolation. Those engaged in forestry are no exception. Modern transport has reduced the disadvantages of living some distance from one's work. There is therefore less objection to forestry personnel of all ranks living in the nearest small town with schools, shops and other facilities which are nowadays taken for granted. In fact the presence of these people adds to the life of the local community, especially if they take an active part in civic, cultural or sporting activities. Moreover, in this way foresters can act as ambassadors of their profession, contribute to a better understanding of forestry and at the same time listen to and learn from the views of others. All this gives forestry a better chance not to be ignored when issues of local or regional policy are decided. Those responsible for forest policy can contribute to these trends by no longer insisting on foresters living in the forest, and by locating forest offices in suitable small towns. Housing that may become vacant by moving personnel to local centres of population may be let as holiday cottages. Examples of this are found in Britain, where the Forestry Commission established so-called forest villages after World War II to house foresters and forest workers for its major afforestation programme. It soon became evident that these people and even more so their families preferred to live in the nearest small town with shops, schools, cinemas etc. Now these forest villages have taken on a new lease of life as holiday homes for tourists. These tourists are doubly welcome because they also provide additional incidental employment and income for some of the families of the forest employees.

There are of course also disadvantages when those engaged in forestry no longer live near their work. The risk of theft is increased and so is damage by visitors; rapid action in the event of fire also becomes more difficult. Fortunately, there will always be some foresters and forest workers who prefer to live right out in the country, especially those with a farming background. They should be encouraged to do so.

Forestry leads to the establishment of forest industries in rural areas where other industrial employment may be scarce. It is very difficult indeed – at any rate in countries with a market economy – to achieve a well balanced geographical distribution of industry; there tends to be a cleft between crowded industrial centres and almost empty rural areas. 'Industry attracts industry' seems to be a true slogan. Agriculture and forestry, however, are bound by their nature to the countryside; together with the industries which are connected with them they are the main sources of employment in the rural space, indeed sometimes almost the only ones except where there is tourism. Forest industries not only offer additional employment opportunities, but work of a different kind from forestry and agriculture; forest industries may bring new life to old towns or give rise to whole new towns with schools, hospitals, civic and sporting centres etc.

In some countries farmers have switched from full time farming to part time farming so as to work part time in industry. These are resident, reliable workers and the money they earn in industry is an inducement not to abandon their farms. As already mentioned elsewhere, part time farming and farm amalgamations are complementary ways of dealing with the reduced manpower requirements in farming which have resulted from mechanization and other improvements in farming methods.

The mechanization of forest operations impinges on rural community development in two ways. The higher the degree of mechanization and the larger the machines, the more specialized are the skills needed and the fewer the jobs. The view is now gaining acceptance that there should be an increased emphasis on what is sometimes termed *intermediate technology*; small machines, better accessories to farm tractors for timber extraction etc. In this way forest work remains less specialized, less capital is tied up in machines, the risk of environmental damage is reduced and there are more jobs.

Tourism has brought prosperity to many rural areas, and forests are among the tourist attractions. Some tourists spend holidays in small hotels or on farms where they can be in contact with the local people and thus get an idea of what farmers or forest workers have do do and how they live. Such tourists are welcome consumers of all kinds of agricultural products and when they come from abroad, they bring in foreign currency. This can be important. In Austria, which has a population of 7.5 million, foreign visitors accounted for over 100 million visitor-nights in 1980. Clearly, tourists spend a lot of money in rural areas; but to attract tourists, investments are necessary in order to bring accommodation both in hotels and private lodgings up to the standards required.

Forestry's contributions to tourism have been discussed under recreation in Chapter 4.2, where it was pointed out that the main emphasis should be on providing opportunities for the peaceful enjoyment of nature: marked tourist paths with benches etc. While forest authorities should encourage tourism, the interest of forestry must be safeguarded. The costs of precautions against fire and litter dumping by tourists should not have to be borne by forest owners. An equally important aim of forest policy must be the prevention of uncontrolled encroachment on the forest area by second residences and installations such as ski pistes.

A little holiday home in a forest is pleasant for the owner and may even add variety to the forest scenery. The trouble is that one little holiday home is followed by ten and then by a hundred, because if building permission is granted to one person, how can others be refused? Similar considerations apply to installations such as ski pistes. These are difficult problems that raise wider issues. One approach that has been suggested is to discourage such developments where they have not yet started and to allow developments to continue subject to suitable planning controls where encroachments have already reached a point of no return. Understandably, this approach is not popular with communities to be excluded from further developments. Why should they be put at a permanent financial disadvantage? Could they be compensated in some way out of the tourist revenues elsewhere? These and related questions raise issues of general social and environmental policies, but the forestry implications must be given due weight when these matters are considered.

Forest ownership also exercises an influence on rural community development. Woodland owners who live in the locality are obviously more closely connected with local affairs than those living farther afield, but absentee owners with an interest in forestry who invest capital in forests, create employment and bring new ideas, may also be beneficial. It is the indifferent woodland owner, whether local or not who contributes least. Forests owned by communes often provide a useful source of finance for public works in addition to providing the members with fuelwood and sometimes also with timber.

In the process of farm rationalization in Western Europe considerable areas of land have been freed for afforestation. This helps to curb agricultural over-production and will eventually reduce the timber deficit, two very desirable objectives. What is less desirable is that these outgoing farmers are sometimes offered insufficient technical and financial assistance to enable those who are interested to undertake the afforestation themselves instead of having to sell out to the state or to others with the necessary capital and know how. Afforestation by the outgoing farmers may be less cost effective, but their interests might be served better if the land did not have to change hands; and these interests should also carry some weight.

Finally, foresters and forestry can also make a direct contribution to the *social life* of rural communities. Cities offer their inhabitants innumerable oppotunities for entertainment: theatres, cinemas, concerts, dance halls, sports stadia and a hundred others. There is no reason why life in the country should be dull and boring; but it is the people themselves who must organize their own entertainments and thus create surroundings 'full of life' and attractive to live in. Of course those responsible for forest policy can only have a limited influence in these matters, but they should give what encouragement they can. A few examples may illustrate the point.

Tree felling and tree planting competitions can be great fun for competitors and onlookers alike and if carefully arranged, they help at the same time to promote fast working to a high standard. Competitions for the best stands to illustrate the results of silvicultural treatments are also popular in some countries and have proved instructive. Forestry and forest machinery exhibitions either on their own or as a part of wider shows are other examples that lend themselves to a good mix of instruction, publicity and entertainment.

To summarize: the main policy message from this section is that governments and forest authorities can best contribute to rural community development in both developed and in developing countries by encouraging their personnel not only to take an interest in forests and trees but also to talk and listen to people and play an active and constructive part in the life of the communities in which they live and work.

References

Sections 5.1 (Protection) None.
Sections 5.2 (Biomass) and *5.3* (Wood and Energy)
FAO (1981) Map of the Fuelwood Situation in the Developing Countries, FAO, Rome.
FAO (1981) Wood Energy. UNASYLVA, Vol. 33, Nos 131 and 133, FAO, Rome. (Special Edition 1 and 2).
Strub, A., Chartier, P. and Schleser, G. (1983) Energy from Biomass, 2nd E.C. Conference, 1982. Applied Science Publishers Ltd.
Sections 5.4 (Tropical Moist Forest)

Boerboom, J.H.A. et al. (1983) Human Impact on Moist Tropical Vegatation (in: Man's Impact on Vegetation), Dr W.Junk, The Hague.

Grainger, A. (1980) The State of the World's Tropical Forests in: The Ecologist, 10: 6-54.

Hallsworth, E.G. (ed.) (1982) Socio-economic Effects and Contraints in Tropical Forest Management. John Wiley & Sons, Chichester.

INTERNATIONAL UNION FOR CONSERVATION OF NATURE, (1980) World Conservation Strategy.

Lanley, J.P. (1982) Tropical Forest Resources, FAO Forestry Paper 30, Rome.

Meyers, N. (1980) Conversion of Tropical Moist Forests. National Academy of Sciences, Washington, D.C.

Meyers, N. (1979) The Sinking Ark, Pergamon Press, Oxford.

Poore, D. (1976) Ecological Guidelines for Development in Tropical Rain Forests, IUCN.

Sommer, A. (1976) Attempt at an Assessment of the Worlds Tropical Forests, UNASYLVA 28/112-113: 5-25.

Spears, J.S. (1979) Can the wet Tropical Forest survive? Commonw. For. Rev. 58: 165-180.

Spears, J.S. (1980) Can farming and forestry coexist in the tropics? in: UNALYSLVA, 32/128: 2-12.

Stoel, Th.B. et al. (1981) Actions Needed to Conserve Tropical Moist Forest, Natural Resources Defence Council, Washington DC.

UNESCO (1978) Tropical Forest Ecosystems. Natural Resources Research 14. UNESCO/UNEP/FAO, Paris.

Ulrich, B. (1981) Bodenchemische und Umwelt-Aspekte der Stabilitat von Waldokosystemen in: Schriften Forstl. Fakultat, Universitat Gottingen, 69, 19-29.

US INTERAGENCY TASK FORCE, (1980) The World's Tropical Forests: A policy, Strategy and Program for the United States. Report to the President, Washington, DC.

Sections 5.5 to 5.7 (Farm forestry, agroforestry, rural community development)

Ben Alam, B. en T Van Dao (1981) Fuelwood production in traditional farming systems: UNASYLVA 33 (131): 13-18.

Bergmann, Th. (1981) Comparative Agricultural Policy. Sociologia Ruralis, vol. XXI (3/4)

Bonvoisin, S. (1982) The implications of community forestry projects. Commonw. For. Rev. 61, 2, pp. 145-150.

Budowski, G. (1982) The socio-economic effects of Forest Management. The lives of people living in the area; the Case of Central America and Some Caribbean Countries. in: E G. Hallsworth (Ed.), Socio-economic Effects and Constraints in Tropical Forest Management. John Wiley & Sons, Chichester

Evans, E. (1982) Plantation Forestry in the Tropics. Oxford University Press.

FAO, (1978) Forestry for Local Community Development. FAO Forestry Paper, No. 7, Rome.

Gilg, A.W. (1978) Countryside Planning. University Paperback 1979, Methuen & Co. Ltd. London.

Gonzalez de las Salas (Ed.), (1979) Proceedings Workshop Agroforestry Systems in Latin America. CATIE, Turrialba, Costa Rica

Grainger, A. (1981) Integrating Farming and Forestry. Reforesting Britain: a Special Report: The Ecologist, Vol. II, No. 2, 70-73.

Green, B.H. (1981) Countryside Conservation. Resource management series 3. George Allen & Unwin, London.

Haynes, C.D. (1978) Land, trees and man (Australia). Commonw. For Rev 57(2)· 99-106.

Huguet, L (1980) L'Association de la Foret et de l'Agriculture dans la Chine agricole, de Beying a Guangzhon. in: Bois et Forets des Tropique, 189.

Kundstadter, P. et al.(1979) Farmers in the Forest; Economic Development and Marginal Agriculture in Northern Thailand. East-West Centre, Honolulu.

Pischke, J.D. von (1981) The political economy of specialized farm credit institutions in low-income countries. World Bank, Washington, DC.

Poulsen, G. (1981) The function of trees in small farmer production systems. Forestry for Local

Community Development Programme, Report GCP/INT/347/SWE, FAO, Rome.

Raintree, J.B. (1982) Readings for a Socially Relevant Agroforestry. International Council for Research in Agroforestry. (I.C.R.A.F.), Nairobi.

Ratna Murdia (1982) Forest development and tribal welfare; analysis of some policy issues. E.G. Hallsworth (Ed.), Socio-economic Effects and Constraints in Tropical Forest Management; John Wiley & Sons Ltd. Chichester, New York.

Ruthenburg, H. (1971) Farming systems in the Tropics. University Press, Oxford.

Spears, J. (1980) Can Farming and Forestry co-exist in the Tropics? UNASYLVA Vol. 32, No. 128: 2-12.

Wiersum, K.F. (Ed.) (1981) Viewpoints on Agroforestry. Agricultural University Wageningen.

6 Institutions and administration

Fred Hummel

Decisions on forest policy are meaningless without the laws, institutions and administrators to implement decisions. It is therefore a matter of policy to ensure that these instruments of implementation are adequate for the purpose. They will be discussed under the following headings: Forest ownership, legislation, taxation and incentives, education and training, research, organization, personnel.

6.1 Forest ownership

Forest ownership varies greatly from country to country, but virtually all countries with any forests at all have state forests and many, including some with centrally planned economies, have private forests. In Yugoslavia, for example, about 30% of all forests are privately owned and in Poland and the German Democratic Republic about 15%. Ownership by local communities (villages, towns etc.), production co-operatives (mainly in countries with centrally planned economies), industries and institutions (e.g. for nature conservation, ecclesiastical bodies, pension funds) is important in some countries. The ownership pattern is usually different from that of agriculture in that more forests tend to be owned by the state.

6.1.1 State forests

State forests are almost universally recognized as an essential component of a country's forest estate. State in this context refers not only to a country as a whole, but also to parts of a country such as the 'states' in the USA, the 'Provinces' in Canada, the 'Länder' in the Federal Republic of Germany, or the 'Cantons' in Switzerland.

The state is in a better position than an individual or even a small community to ensure the long term continuity so essential to forestry, and to translate forest policy into action. In countries without a forestry tradition it is virtually only the state on its own land that can get an effective forestry programme started. It also appears to be widely accepted that government forest officers with personal experience of forest management have an advantage over those without such experience in understanding the management problems of woodlands in other ownership. As state forests are usually in large units, they provide opportunities for economies of scale and for exercising a stabilizing effect on markets, e.g. sales can be reduced when demand is poor and increased when demand picks up. The scope of such countercyclical action is, however, limited by various constraints such as the need to provide stable employment to forest workers and by the facts that wood is traded internationally on a very large scale and that there are many wood products.

Ownership by the state or other public authority is also an advantage where protection of the environment or recreation are important, because in these cases most of the benefits go to the community at large and not only to the forest owner.

Against all these advantages of state forests must be set some possible disadvantages. On the one hand, the dead hand of bureaucracy may stifle initiative; on the other hand, the very ease with which new ideas may be implemented in state forests has led in some countries to somewhat rash innovations on a large scale which soon were proved unsound. While the general standard of management is high in state forests, there is less scope for integrated land use and for adapting management to local circumstances than in the woodlands belonging to a farmer or a local community. It is probably also true, that even allowing for environmental and other constraints, state forests are rarely managed as profitably as they could be; but this can largely be remedied by providing a suitable organizational structure for state forests (see 6.6) and by the appropriate training of forest officers (see 6.4).

6.1.2 Forests owned by individuals

In some countries, especially those with a long forestry tradition, a large proportion of the forests are owned by individuals. In contrast to agriculture, few woodland owners, whether large or small, depend on their forests as their main source of income; this may have implications for forest management and forest policy. The average size of private forest holding is usually small, even in countries with some large privately owned forest estates. Thus, in the European Community some 19 million ha of forests in private ownership are shared among 3 million owners, some of whom, however, own several thousand ha. These large estates are generally well managed by professional forest officers.

From the point of view of good forest management, it may be of little consequence whether well managed large forest estates are owned by individuals, the state or some other body, but any taxation or other policy measures which lead to the splitting up of well managed estates are almost bound to be against the national interest, because fragmentation renders efficient management more difficult.

Indeed it is the problems associated with fragmentation that constitute the main policy issues of private forestry. Forestry requires special knowledge and skills which few owners of small woodlands possess or find it worth their while to acquire, given that forestry can only add marginally to their income. Where forests have been over-expploited in the past or neglected, there is also the problem that matters can only be put right by a considerable injection of capital, from which there may be no return for a long time. In spite of all these difficulties much is to be said in favour of small, privately owned woodlands. Valuable new developments in silviculture have been pioneered by owners, with the necessary time, intelligence, knowledge and enthusiasm. Some of the methods evolved on such estates have proved to be of wide practical application. In Britain, for example, the introduction of exotic tree species and the afforestation of peat sites was largely pioneered by private woodland owners. We must also recall that the economic production of timber is not the only objective of forestry and that much broader issues are at stake. The close integration of farming and forestry which benefits the farm (see Chapter 5.5) and contributes to the beauty of the landscape and the conservation of wildlife is best assured if farmers own woodlands. Experience also suggests that widely distributed ownership of land makes for political stability and that ownership of woodlands may help to foster a better understanding and interest in forestry among citizens.

The disadvantages of fragmentation may be reduced if the management of woodlands belonging to several owners can be co-ordinated e.g. by associations of woodland owners of by other enterprises which undertake the harvesting and sale of timber or even the entire management of woodlands. The success of such approaches depends very largely on the competence and enthusiasm of all concerned. The main difficulty is that woodland owners tend to be very independent minded.

Measures to promote the amalgamation of small woodlands are advisable if the laws and customs of inheritance in a region have led to the successive splintering of landed property over centuries among descendents, many of whom may have emigrated and cannot even be traced. In this situation, management may be impossible without prior land consolidation. In countries where land consolidation measures require the approval of a majority of the owners concerned, the only practicable way of achieving such a majority may be to regard abstentions as votes in favour, as has been done in parts of Switzerland.

Woodlands are sometimes in the joint ownership of groups of individuals or families. In Austria and Switzerland, for example, there are substatial areas of forest which are owned by groups of residents in a village or small town; ownership is linked to rules of residence and status often going back for centuries;

the rules are such that it is not easy either to relinquish or acquire a share in such ownership. These forests differ from communal forests because they are *not* owned by the village or town as a whole and they are *not* managed by the local authority: they are owned and managed by the group of individuals concerned. The system is said to work well where it exists, but whether its introduction elsewhere whould be advantageous appears doubtful.

Quite a different form of joint ownership has recently been developed, notably in Britain: so-called 'forestry investment companies' purchase woodlands and offer shares to private investors. The ownership is vested in a trust belonging to the shareholders; the company arranges for the management of the woodlands; the costs of forest operations and management and the proceeds from the sale of produce are divided among the shareholders in proportion to the number of shares held. The system combines the possibility of relatively small investments in forestry with units of forest management which are sufficiently large to permit rational working. The fact that changes in the ownership of shares do not disrupt continuity of management, also reduces some of the disadvantages generally associated with absentee ownership. Whether or not such a system of joint ownership is likely to be beneficial and successful must depend on the particular circumstances of a country.

6.1.3 Communal forests

In several countries of Western Europe, forests belonging to local village communities constitute a significant proportion of the country's forest estate and have been of great benefit to the communities concerned, supplying them with produce and furnishing revenue which helps to defray the costs of public works. The difficulties associated with collective ownership must however not be underestimated. Holopainen (1981) referring to Finland, states:

'An endeavour has been made to establish *forests in collective ownership* mainly for the purpose of avoiding the drawbacks associated with small-scale ownership. These attempts have met with only minor success.'

Success depends largely on being able to secure a willingness to co-operate on matters such as road building and marketing, but this co-operation is often not easy to achieve, because of an unwillingness to sacrifice complete independence and freedom of action. Also in some developing countries, village forests have been introduced over a long period with varying degrees of success. Troup (1940) states:

'In agricultural countries like India and many of the Colonies, forests set apart for village requirements are a matterof great importance in the rural economy of the country. The produce yielded includes fuel, poles and withes for house building, wood for agricultural and domestic implements, fencing and thatching material, fodder, grass and other minor products, including edible fruits and plants which are particularly useful in time of famine. Under certain conditions regulated grazing may be added to the benefits afforded by such forests, particularly in famine years.'

Also, according to Troup (1940) the Forest Department in the Punjab began already in 1866 to establish plantations primarily to meet the needs of the local agricultural populations. Similar developments followed somewhat later in other developing countries.

The welcome emphasis placed on the concept of 'Forests for the People' by the World Forestry Congress at Jakarta in 1978, has given a new impetus to the establishment and management of village forests as well as to agroforestry and the role of forestry in community development, two subjects which are dealt with in more detail in 5.6 and 5.7.

6.1.4 Co-operative forests

Co-operative forests exist mainly in countries with centrally planned economies, but their importance varies. They are common in the German Democratic Republic (GDR) and in Hungary where they account for 20-25% of the total forest area.

In Bulgaria, Czechoslovakia, Romania and the USSR the percentage lies between 2 and 8 while in Poland this category of wonership is almost completely absent.

The co-operative forests are managed by the co-operatives in the same way in all the centrally planned economies, but the ownership differs somewhat from country to country for historis reasons.

In the USSR private ownership of land was abolished in 1917 when all land was nationalized. The use of cultivated areas was transferred to the peasants free of charge. Included in this transfer were the limited areas of forest already used by the peasants. After the collectivization of agriculture, these privately used forests became part of the land granted by the State to the collective farms for use in perpetuity. The ownership of the land, however, remains in the hands of the State. These forests are intended to serve the needs of the co-operatives and the personal needs of their members.

In the other Eastern European countries, the ownership of the land occupied by co-operatives varies somewhat, but it has not been nationalized. In *Hungary* for example, part of the forests managed by the co-operative farms are owned by these farms while another part is owned by the members of the co-operatives. The tendency is for all these forests gradually to become owned by the co-operatives.

These forests cover in the first place the needs of the co-operative farms and their members but they also produce some forest products for sale. In the *GDR* most of the land now managed by the co-operatives had first been granted to the peasants through land reform. In *Czechoslovakia* farm forests became an integral part of co-operative farms during the process of collectivization. It is, however, worth noting that the area of co-operative forests culminated in 1963 and has been declining since then because the state forest administration has gradually been taking over the management of these forests. In *Poland*, the area of co-operative forests is negligible, because the agricultural and forest land has not been collectivized and most of it continues to be owned and managed by the individual farmers.

6.1.5 Forests owned by industry

Forest industries own substantial areas of forest in several countries, for example Brazil, Finland, New Zealand, Sweden and the USA. The general standard of management in these forests is high and the benefits to the industries concerned are obvious; they have at least part of their raw material supply assured and they can manage their forests so as to best serve their industrial objectives. Up to a point, the efficient management promoted by this close integration of forest and industry may also be in the national interest, but both Sweden and Finland, two of the leading countries in forestry and forest industry, have found it necessary for socio-economic reasons to restrict the further purchase of forests by forest industries. Holopainen (1981) explains:

> 'The transfer of land from private to company ownership resulted in a number of social problems. For instance, the owners of the holdings often remained as tenants on the farms bought by companies or were entirely dispossessed. Hence the growth of land ownership by companies was criticized and a law was eventually issued in 1915 under which companies engaged in woodworking industry and the timber trade were allowed to purchase only such forest land that was not considered desirable for agriculture.'

These restrictions were reinforced by subsequent legislation. The principal motivation was to prevent rural depopulation by securing the well-being of agriculture and the farm population. Large scale forest ownership by forest industries may also give these an inordinate influence over the price of the wood they buy; and it may accentuate the imbalance in bargaining strength that may in any case exist when a few large and powerful industries purchase supplies from a large number of small woodland owners.

Moreover, forest ownership by industries may lead to wood not being put to the best use. It may, for example, be in the interests of a pulp mill to convert into pulp wood which is suitable for sawing although the latter use might be in the broader national interest. This question is less likely to arise if the industry owning the woodland has its own pulp mill, saw mill, particle board mill etc.

Similar considerations to those of woodland ownership by forest industries apply where an industry has a concession to manage and exploit an area of forest belonging to someone else.

6.1.6 Forests owned by inbstitutions

Institutional owners of forests range from ecclesiastical bodies to pension funds and from organizations concerned with nature conservation to private investment companies. In spite of these contrasts, most of these institutional forest owners have two characteristics in common; first, forestry is rarely a primary objective but rather a means towards achieving the institution's main objectives; and secondly, since institutions tend to outlive individuals, there may be a greater likelihood of continuity than in the case of ownership by an individual. In a few instances, governments have restricted such ownership. Thus, ecclesiastical ownership in Finland was treated in a similar way to industrial ownership and already early in the 19th century in Bavaria, forests belonging to ecclesiastical bodies were nationalized, together with other Church property when it was felt that the Church had become too rich and powerful.

There can, however, be little doubt that institutional owners of all kinds may make a useful, if modest, contribution to the development of a country's forest estate. Ownership by organizations concerned with the conservation of nature is to be particularly welcomed because it associates individuals more closely with conservation than if the forests in question are owned and managed by the State. In some countries such as Britain, it has also been found that private owners of woodlands of special environmental interest are often willing to donate them to a private organization, such as the National Trust, when they would be unwilling to donate them to the State.

Ownership by pension funds, insurance companies, banks and private investment trusts is a relatively new development in some industrialized countries. The motive is the search for an investment which is relatively safe, does not get eroded by inflation and achieves long term capital growth. In some cases, there may also be tax advantages. The dangers inherent in this type of absentee land ownership are offset by the capital resources and management know how thus injected into forestry.

6.1.7 Policy considerations

Good forest management and bad forest management are possible and have occurred under every type of ownership, although some types may be more conducive to good management than others. Where for any reason connected with forestry or for broader political reasons, some change in the existing balance of ownership is made, gradual action over a long period which permits adjustments to be made in the light of experiece is usually preferable to sudden, drastic measures which could disrupt management and may have to be withdrawn or reversed if they fail to achieve their purpose.

All categories of forest ownership can contribute to the development of forestry by fulfilling complementary roles which none can fulful alone. Measures to influence the pattern of ownership may be necessary in special cases, such as that cited from Finland, but usually it is preferable to solve problems that may arise in other ways, e.g. by incentives or by strengthening supervision by the State.

Woodland owners in most countries of the world with a forest administration are in fact subject to some control and supervision by the State. There are very good reasons for this. In protection forests, mismanagement may cause harm elsewhere; in production forests any profit to the present owner from over-exploitation, will be at the expense of future generations. Because of the impact of forest management on the general welfare of the community at large, it is not illogical that Switzerland, one of the most democratic countries in the world, is also among those with the strictest state control over forests in other ownership.

There are broadly three methods by which the State may exercise its influence: Restrictions and obligations, state assistance, management by the State.

Restrictions and obligations which are imposed on forest owners in various parts of the world include the following:

1 the right to acquire forests restricted to people with some forestry experience;
2 restrictions on the conversion of forests to other land use;
3 restrictions on the choice of species and of plant reproductive material (seed, plants, vegetatively propagated material); these restrictions may be for phytosanitary reasons, for the purpose of landscape conservation, or to prevent the propagation of genetically inferior trees;
4 various restrictions on fellings, e.g. limits concerning the volume of timber and/or size of tree which may be felled; the prohibition or limitation of clear-fellings as a means of preventing erosion on hillsides; the prohibition or restriction of fellings for some years after the purchase of a forest in order to discourage land speculation; restrictions of fellings for reasons of landscape conservation;
5 requirements as to silvicultural practice, e.g. control of grazing, precautions against fire and disease:
6 measures to secure proper management: compulsory plans of management which have to be approved by the forest administration; in some countries,

private forests above a certain area must be managed by a professionally qualified forest manager.

As with all restrictions and prohibitions, they are not worth introducing unless they are both necessary and enforceable.

Various kinds of *state assistance* complement restrictions and obligations. In fact, as far as practicable, incentives are preferable to restrictions. Responsible woodland owners respond more readily to the carrot than to the stick, but the stick must be held in reserve for those who are less willing to co-operate. It may sometimes be convenient to link an obligation with an incentive, for example, by making financial assistance conditional on working to an approved plan of management.

Assitance by the State may be financial, technical or organizational. The financial aspect is dealt with in section 6.3 Taxation and Incentives. Technical assistance generally takes the form of professional advice, assistance in the preparation of management plans, circulation of information on markets and other relevant topics, and help with the marketing or produce, e.g. by allowing timber from private woodlands to be included in auction sales organized by the State. One of the most common forms of organizational assistance is in the formation of groupings of small forest owners. The formation of such groupings may be encouraged and administration simplified if State assistance is channelled through such organizations. The promotion of co-operatives has however, often proved a failure. Success seems to depend more on the people directly responsible than on the organizational details.

Whatever assistance measures may be deemed appropriate, the basic long term policy objective should always be to help forest owners to help themselves.

In some instances the State manages forests owned by others. Management by the State is, for example, obligatory for the communal forests in France and Belgium and for the privately owned farm forests in Yugoslavia. There are also instances where this system is operated by voluntary agreement between the state and the owner. The system works well where the state forest officers have succeeded in establishing a spirit of friendly and close collaboration with the woodland owners by showing an understanding of their problems and by taking account of their reasonable wishes.

6.2 Legislation

Most laws of a country apply to forests and the activities relating to them in the same way as they apply to any other activities. These laws are not directly related to forest policy but some may profoundly affect its scope. Legislation concerning tenure and use of land is a case in point.

The development of the forest resource may require changes in general legislation or at least some exemptions from and additions to it. In addition, there are matters that inevitably require specific forest legislation. The legal aspects of forestry are highly complex and all that can be attempted here is to focus attention on some major issues of policy concern.

6.2.1 General considerations

Forest legislation, in the minds of many, is primarily associated with preventing the misuse of the forest and protecting the rights of individuals and the general public (e.g. in matters of access); but legislation should also serve as an agent for development and as an incentive to good management (e.g. through tax provisions).

The form and content of legislation affecting forests depends largely on the general legal system of a country; some prefer a very detailed and rigid body of law, while others prefer to have fewer and simpler laws and to give correspondingly more powers of decision to their administrators. Detailed laws promote equality before the law because less scope is left for different interpretations in different parts of a country, but detailed law may complicate administration and thus impair efficiency. Moreover, the inclusion of technicalities, which may be made irrelevant by technical progress may lead to the need for frequent revisions. Giving more discretion to administrators opens certain doors to corruption, but complex legal procedures open others. For example an elaborate system of granting permits to fell timber makes it just as easy for officials to exact bribes as a simpler system. The remedy for corruption lies mainly in creating job satisfaction and providing reasonable levels of remuneration; but the penalties for corruption must also be severe.

A two tier system of legislation is a useful way of differentiating between basic legislation which rarely needs revision and subordinate legislation which may need to be changed more frequently. The parent law is framed in such a way that its provisions are skeletal but authorize in clear terms a well defined authority to enact subsidiary legislation on specified topics and in specific circumstances. Such two tier systems are common in countries with a federal structure. Recent examples for the forestry sector are the new forest law of the USSR of 1st January 1978, and the federal forest law of the Federal Republic of Germany of 7th May 1975. Even where there is no federal structure, two tier legislation has been found useful for certain forestry purposes, especially where it is desirable to adapt the law to local conditions. For example general laws on forest protection or the recreational use of forests, may include enabling legislation for the issue of specific regulations and rules for protective measures and recreational facilities in particular areas.

Whatever the legislative traditions of a country, a law is worse than useless if it cannot be enforced, because in that case it will be ignored; and if one law is treated with contempt, the respect for other laws will also be undermined. For a law to be enforceable it does noet suffice to have adequate arrangements to detect and deal with contraventions, although these arrangements are of course essential; it is equally, if not more important, for a law to have some measure of public support, which in turn presupposes that its reasons and purpose are understood and that it does not impinge on any basic necessities of life. This may sound obvious, but there are many examples to show that these rules are not always observed. Thus insufficient effort is sometimes made to explain to the public why access to forests may have to be restricted for fire prevention, or for the conservation of fauna and flora or because felling or road building operations could make it dangerous for visitors to be in the forest. Worse still are felling prohibitions and the prohibition of the clearing of forest for cultivation if the people affected have no other means of subsisting. Unfortunately, it is sometimes easier for politicians to introduce such prohibitions than to initiate the necessary agricultural and forestry developments which will no longer make it necessary for shifting cultivators to destroy the forest.

For the enforcement of forest laws, government forest officers and foresters in some countries are given certain police responsibilities, including the power of arrest. Where these police responsibilities exist, the personnel concerned normally wear uniform to facilitate identification. In other countries the staff of the forest administration has no special police function. There is a fundamental difference in approach here. In the one case government forest personnel are seen essentially as an extension of the arm of the law, while in the other the emphasis is on their being servants of the community who report contraventions of the law to the police. Countries which have not yet reached a decision on this issue, would do well first to study how each system works.

Some laws concerning forests are incorporated as special provisions in general legislation; examples are forestry sections on taxation, nature conservation or plant health legislation. Laws, however, which are more specifically forest orientated, e.g. those concerned with forest protection, management and exploitation are usually brought together in a separate forestry statute or forest law. There are good reasons for this: a separate law is easier to draft, easier to handle by those concerned with its enforcement and easier to amend.

6.2.2 Drafting and reviewing forest legislation

The first point to make is that in order to reconcile the technical requirements of content with an appropriate legal formulation, technical and legal experts must collaborate in the drafting of forest legislation and it should be the aim of both parties to express the laws in language which people can understand.

The revision of an existing forest law or the drafting of new forest legislation must be preceded by the collection and analysis of several types of data. King (1969) states:

'All laws relevant to forestry ought to be examined: land law, land-tenure systems, rules of succession, various taxation and industrial laws, the law of contract and the legal conditions under which labour may be employed, minimum wage rates, etc. The criminal law should also be carefully scrutinized, for any recommendations with regard to such matters as penalties would have to fit within its framework. Most important of all is constitutional law. It is often extremely difficult to alter the terms of a nation's constitution, and it is therefore essential that the full implications of its clauses be clearly understood before legislation for any form of activity is drawn up.

Legal investigations are, in themselves, not enough, however. If the laws that are finally formulated for the forestry and forest industries sector are to have the desired propulsive effect, if they are to be employed as an instrument of development it is equally advisable to consider any land capability classifications that are available, any land-use plans which the nation may have and, indeed the general socioeconomic situation.

With this background knowledge, the nation will then be in a position to revise those other laws which it considers might hinder forestry development.'

Any such comprehensive review of the situation will act as a guide and will reveal weaknesses to be rectified. Speaking of developing countries King (1969) lists twelve common defects, some of which at least are also to be found in developed countries. They are:

1   inconsistencies in the general forest law;
2   confusion in the definition of forest land;
3   imprecision in the definition of the powers of forest officers;
4   conflict between forest legislation and other related laws;
5   incompatibility of forest laws among the states, and between them and the central government in federal types of government;
6   multiplicity of legally sanctioned authorities concerned with forestry matters even in some states with a unitary form of government;
7   dual control of forest lands;
8   prevalence of various rights of usage over forest lands;
9   absence of control over privately owned forests;
10 unnecessary number of types of forest tenure;
11 unsuitability of existing types of forest tenure for development;
12 inadequacy of legislation affecting investment in forestry and forest industries.

6.2.3 Specific legal issues (land, production, protection)

There are many forestry matters that may require legislation ranging from taxation to the quality of forest reproductive material and from the rights and duties of forest officers to wild life management. In order to decide whether or not legislation is needed, it is first necessary to be clear about the policy objectives to be achieved, and the alternative or complementary methods available for the purpose. In the present context we shall only consider land, production and protection.

Land

Important legal distinctions with policy implications exist between ownership, occupation and rights by third parties. For example, ownership of a forest may be vested in the state, but a local community may be legally entitled to manage it, and certain individuals, usually of the same community, may be entitled to wood from it free of charge for their personal use. These rights of occupation and of use usually arise when ancient practices are confirmed by law. Where such rights become an obstacle to efficient forest management, as they very often do, steps should be taken to try to extinguish these rights in a way which is just and has the agreement of all concerned, which may of course prove difficult. The methods that have been adopted include compensation in cash or granting the ownership of part of the forest to the right holders.

The pressures on the forest are continuing to increase. Adequate legal safe-guards against any unreasonable conversion of forest to other use – roads, ski pistes, building holiday homes etc – are therefore essential. Among the strictest laws are those of Switzerland and Austria, where the conversion to other use of any land classed as forest may be made conditional on the afforestation of a similar area of bare land elsewhere. In many countries any forest area which is felled must be reforested within a specified time limit.

In less developed countries without adequate forest administrations, such laws to protect the forest estate are difficult to enforce, but the prevention of forest destruction is just as important. In some countries, including those under former British rule, forests which were considered important either for the production of timber or for environmental reasons, were legally gazetted as permanent forest reserves. The system of forest reserves has been criticized in some quarters as having made forest administrations more intent on 'defending their territory' than on actively managing the forests for the benefit of the local people. While this criticism may not be unfounded, the system has undoubtedly enabled many forests to be brought under proper management and it has prevented some of the forest devastation that has been going on elsewhere. Such a system at least provided the opportunity for careful planning before major decisions on changes in land use are taken.

The question of rights of access to forest land by the general public has become an important issue in developed countries with a dense population or with a highly developed tourist industry. In developing countries the problem usually arises only at tourist centres because the people living on the land do not need to walk in forests to get exercise and fresh air, and only few people in towns have the time and money to escape into the country for pleasure.

The legal response to the problem of access has varied. Taking Europe as an example, in Austria, the Federal Republic of Germany, Switzerland and the Nordic countries, everybody has the right of access on foot to all forests subject to a few minor and well defined exceptions; in Great Britain and the Republic of Ireland at the other extreme, forest owners are entitled to prohibit access, but in practice access is permitted to most state forests.

Perhaps the most important lesson to be learnt from past experience is that countries should decide on a policy concerning access and the recreational use of forests and introduce any necessary legislation *before* the number of visitors creates problems. Where access is granted by law, this should also provide for appropriate exceptions, e.g. for the preservation of fauna, and include rules of conduct concerning such matters as litter, noise, and the use of fire.

Production

In many countries with long established forest administrations, the concept of sustained yield is anchored in legislation in various ways. The object is to ensure that over a period of years, fellings in each forest holding should not exceed increment. Other developed countries such as Britain, which have seen no need for any such legislation, nevertheless tend to have a system of felling permits; as a safeguard against gross over-exploitation and fellings which are environmentally undesirable, these permits are normally issued subject to the condition that the area to be felled be reforested within a specified period. In countries where a major part of the forest recource is as yet unexploited or where most of the forest area consists of immature plantations resulting from recent afforestations, the concept of sustained yield may be a long term policy objective but legislation could do little to help in its achievement and would be irrelevant. But even under these conditions, legal safeguards against over-exploitation are essential.

Where forest owners do not themselves undertake the harvesting of timber but leave this to be done by the purchaser the conditions that are to be observed by the purchaser must be backed by law and a forest authority to enforce the law. This applies particularly where the purchasers are powerful industries, and even more where concessions are granted for the exploitation and management of a tract of forest land over a considerable period of time. The conditions should be simple and clear, and framed in a way which guarantees the future productivity of the forest and provides an incentive to good management. Those who contemplate embarking on schemes of this kind may avoid costly mistakes if they seek the advice of people with experience in these matters. The conditions will usually refer to the rate of exploitation, management practices, environmental constraints (e.g. size of clear fellings), provision for reviews of price (or royalty) and for termination of the concession. General guidance on these points is of little use because what may be sensible under one set of circumstances may be quite inappropriate under others.

Legally binding restrictions on harvesting may have to be imposed in the interests of nature conservation. For example, clear fellings may have to be prohibited on steep slopes since they may cause erosion, or the uses of pesticides and fertilizers should be restricted because they may pollute soil and water. For some purposes, however, positive incentives are a useful alternative to legislation. Planting grants for species desired for environmental reasons are a case in point.

Protection

In Section 5.1 the main policy aspects of forest protection have been indicated with an emphasis on: prevention is better than cure. That section also demonstrates that good management is the surest way of protecting forests in the long term. Legislation can do little to promote good management. Legislation governing the conversion of forest land to other uses has been dealt with earlier in this section. Other aspects requiring legislation are the prevention of fires and of damage through pests and diseases. Legal sanctions are particularly necessary here because failure by woodland owners to take precautions may result in damage elsewhere. In some instances, legislation must not be confined to what happens in the forest. Fires spread from neighbouring farmland and tree diseases may be carried in logs or plants imported from abroad. The devastation caused in Western Europe in recent years by the Dutch Elm disease, a virulent strain of which was imported from the USA in elm logs, illustrates the point. Plant health legislation raises difficult technical problems, for example, import restrictions can only be enforced if the customs officers know how to recognise the material in question. Effective legislation may also depend on international agreement on methods of preventive treatment of timber or plants to be exported and of reliable certification of such treatment. The safest policy is to take no chances; at the same time it is necessary to recognise that vested interests may exaggerate health hazards of imports in order to reduce unwelcome competition for local timber and nursery plants.

6.3 Taxation and incentives

Taxation can be and incentives are means to persuade woodland owners to act in line with government forest policy. As we shall see, taxation is generally a blunter instrument of forest policy than incentives.

6.3.1 Taxation

The main object of taxation is to raise revenue and the way taxes are raised will depend largely on a country's general political, economic and social objectives. Taxation is rarely designed to promote particular forestry objectives because, as a matter of principle as well as for practical reasons, the pattern of taxation is usually similar throughout the economy; that of course does not exclude some modifications in detail for specific sectors such as forestry. From the point of view of forest policy, the main criterion by which to judge taxation is whether or not it provides an incentive to good management. Special characteristics of forestry which are important when considering taxation include the following:

- annual income and expenditure are generally very low in relation to the capital employed in land and growing stock;
- the costs of management are usually all borne by the owner while many benefits – especially the environmental benefits which are difficult to quantify – acrrue to the community at large;
- the long production cycle of up to 100 years or even more may result in an interval of more than one generation between the time when expenditure is incurred and the time when the income is received;
- it is difficult to determine with any degree of precision whether the harvest in any one year or even over a period of years represents income alone or income as well as some liquidation of capital, because trees are both the means of production (capital) and the product (income).

For these reasons, sophisticated systems of forestry taxation are only practicable in countries where forestry is very highly developed; but even here, complexity does not necessarily increase fairness and accuracy because the elements upon which a tax assessment is based cannot by their very nature be determined with precision. Moreover the possible revenue from forest taxes is relatively low in most countries and would not justify excessive administrative expenditure on its collection. Much is therefore to be said for keeping forestry taxation simple. Where the income tax progression is high, an assessment based on actual annual income may inhibit rational management. Owners will have little incentive to adapt supplies to market conditions or even to harvest the whole of the sustainable yield over a period of years. This may be one reason (but not the only one) for the accumulation of surplus growing stock in some countries.

Taxes which are relevant to forest policy generally include some or all of the following:

- income tax which is supplemented in some countries by a capital or wealth tax;
- taxes on land and other real property;
- inheritance and capital transfer taxes.

In addition, there are taxes such as a value added tax on forest and timber products, and various indirect payments such as employers' contribution to social security funds, discussion of which would be beyond the scope of this book.

Income tax

The difficulty of assessing income has been dealt with in various ways. At one extreme, there are countries such as Ireland and the Netherlands with virtually no tax on forestry income; the logical corollary to that is that where losses occur they cannot be set against income from other sources for tax purposes. At the other extreme, there are countries such as the Federal Republic of Germany where an attempt is made to assess the actual annual income; in order to do so it is necessary to differentiate between 'normal' fellings (calculated on principles of sustained yield) and 'abnormal' yields which again are subdivided into 'delayed' yields (if fellings in previous years have been below 'normal'), yields due to calamities (storm, insect damage, etc.) and 'abnormal' yields justified by economic circumstances. To arrive at the net income, the various fixed and variable costs are allocated between these different categories of yield. Most countries have opted for some intermediate system in which the basis of assessment of income tax is an estimate of average income or potential income related to site quality. In Britain, woodland owners virtually have the choice whether all receipts and expenses are classed as income for tax purposes or as capital. From the point of view of promoting good management a system based on site potential rather than on actual income, but with some exemptions for young stands which can provide no income, looks attractive: it favours those who make best use of the productive capacity of a site and penalizes those who do not and who may even be induced to sell out. In practice, however, it has proved very difficult to devise a system of this kind which is simple enough to administer and, at the same time, fair.

Capital taxes

A very modest capital or *wealth tax* at a level lower than net income, may be regarded as a surtax on income, but if the wealth tax necessitates an encroachment into capital, as is now the case in Denmark, it may force owners to reduce the growing stock below the level needed to sustain future production; or it may lead to the sale or gradual dismemberment of a woodland property. As the sustainable ananual yield of timber is nearly always only a very small percentage of the volume of the standing timber which is needed to produce this yield, even a relatively modest level of wealth tax is likely to convert net income into a net loss. It is indeed difficult to envisage circumstances under which it is in the national interest to apply a wealth tax to forest property. More generally, redistribution of forest wealth through taxation of any kind is likely to impair the efficiency of forest management. There are less damaging ways of achieving such an objective.

Capital taxes on land and other property are generally levied on forest estates in the same way as on agricultural holdings. Provided that they are low, as they virtually always are, they present no special problem. A capital gains tax, however, does present a major problem because any increase in the volume and value of the standing timber over a period of years will largely consist of income over a period of years which has been allowed to accumulate in the forest. Britain solved this problem by applying capital gains tax to forest land but exempting standing timber.

Inheritance and capital transfer (gift) taxes also have important implications for forestry. If they are high, they may lead to overcutting and/or to a splitting up of estates into units which are too small to be managed economically. A useful way to enable orderly management to continue after such a change of ownership is to permit tax payment to be deferred or made by instalments as timber is felled in the normal course of management. In Britain, there is the further provision that if a second death intervenes any tax still outstanding from the first death is cancelled. Reduced rates of inheritance taxes have also been tried, but such a system may lead to timber being allowed to accumulate unnessarily in the forest beyond maturity instead of being put on the market; it may even lead to woodlands being bought by rich people for the express purpose of reducing tax liability on death. The misuse of this type of concession is reduced in some countries by restricting the concession to transfers between close relatives and by making it a condition that the property must have been acquired at least some years before the transfer and must be retained by the new owner for another period thereafter.

6.3.2 Incentives

Incentives are a much more flexible and sharper instrument of forestry policy than taxes. They can be specific not only to forestry but to specific activities within forestry, which a government wishes to promote; and the legislative and administrative procedures for introducing, modifying and abolishing incentives are generally much simpler. Even so, changes should not be too frequent or made without good reason if the confidence of the recipients is to be retained, upon whom the effectiveness of such measures ultimately depends. When temporary incentives are offered in order to get something started, their temporary nature should be disclosed at the outset.

In contrast to taxes, however, incentives normally require a specific budgetary provision while the cost of tax concessions, although also very real, are less visible.

Incentives fall into three main categories:

- monetary grants and low interest loans;
- grants in kind (supply of free plants etc.);

- advisory services and other forms of technical assistance.

The following are examples of the purpose for which incentives are given or have been given in some parts of the world:

- afforestation of non-forest land;
- replanting after felling (normally, owners should be expected to bear the full cost of replanting themselves, because they can do so out of the proceeds from the timber sold;
- the planting of trees outside the forest;
- measures such as thinning, drainage, application of fertilizers;
- the conversion of unproductive forests into productive forests;
- the construction of forest roads and tracks;
- procurement of specialized forestry equipment;
- protection against fire and disease;
- making good the damage from major calamities (fire, wind, disease, etc.);
- the formation of forestry associations;
- grants to contribute to cost of skilled management;
- training courses for personnel at all levels.

The main justification for incentives is that the benefits of forestry which accrue to the community as a whole rather than to the owner alone should also be paid for by the community. In state forests the community pays in any case; incentives at an appropriate level to other forest owners therefore merely prevents discrimination against these owners. Where constraints on forest management in the public interest are considered to be an obligation of ownership as such, the resulting costs have to be borne entirely by the owner; felling restrictions on steep slopes to reduce erosion is a case in point in some countries.

In most instances, the cost of incentives is borne by government or other public authority, but some are financed wholly or in part from within the forestry industry itself. The Fond Forestier National in France which provides very low interest loans and other assistance to woodland owners illustrates the point.

As incentives can and should be specific, it is very important to be clear about the objectives, the hoped for results and the priorities. Only in this way can waste be avoided, progress monitored and appropriate adjustments made in the light of experience. Where practicable, incentives should be distinguished from payments for services rendered or costs incurred, e.g. in connection with recreational facilities, or the preservation of habitats for wildlife.

Before deciding what incentives to give, answers to the following questions should be sought:

- which incentives will be most cost-effective?
- how can woodland owners best be persuaded to do what government wants done?

— which incentives will render forests more productive and thus less dependent on aid in the future? Grants for capital investments such as afforestation or forest road construction fall into this category.

In order to ensure that incentives fulfil their purpose, the government is bound to attach conditions which restrict the recipient's freedom of action. The higher the level of incentives, the stricter the conditions are likely to be and the greater that loss of freedom. This is a point which woodland owners sometimes overlook when they ask for more generous financial support from the state.

Incentives in kind instead of in cash, may be preferable under some circumstances, and for certain purposes, but they may also lead to difficulties. For example, the supply of plants for afforestation at low cost helps to ensure the use of suitable planting stock, but may also disrupt the legitimate trade of commercial nurseries where they exist. In this particular case, a solution might be to use these commercial nurseries as contractors to raise the plants in question. A form of aid in kind which is particularly beneficial and almost universally welcome is sound, practical advice given to woodland owners by the staff of the government forest authority. It also makes for better relations if the officials who visit estates to monitor compliance with legislation or progress of grant aided work can at the same time help with constructive advice. In view of the interdependence of agriculture and forestry, those who give technical advice on forestry should have some knowledge of agriculture and vice versa. In an ideal world, they might be the same persons.

Human nature being what it is, aid that is too easy to get is rarely appreciated and often wasted; and initial enthusiasm may wilt. All over the world there are examples of tree planting schemes which have failed either through carelessness in the handling of plants at the time of planting or more often through lack of subsequent tending. In order to reduce these risks, the recipients of incentives in kind should normally be required to make an appropriate contribution in cash or labour, and there must be adequate monitoring of progress accompanied by measures which will sustain interest and improve knowledge such as courses, practical exercises, and competitions.

Because incentives are specific, they may conflict with other policies unless they are carefully formulated. Thus grants for afforestation have encouraged the planting of sites which were intended for preservation in their existing condition as habitats for rare species of wildlife or simply as open spaces with a beautiful view. Conversely, forestry has suffered through the granting of incentives for other purposes. In the EEC for example, the agricultural subsidies which have led to a surplus of some farm products have discouraged the afforestation of land which is submarginal for farming but would be highly productive for the growing of timber, which the Community has to import in large quantities. Foresters can only hope for a sensible resolution of such conflicts of interests if they make their voice heard and, at the same time, display an understanding of the other interests involved.

6.4 Education and training

The success of forestry depends in the long term very largely on the skills and motivation of the men and women engaged in it; education and training are therefore of fundamental importance. We are here concerned primarily with training i.e. the preparation for a specific career rather than the general strengthening of the powers of body and mind implied by the word *education*, but a sound education is of course essential before training in particular skills can begin. The question of spreading a knowledge of forestry among the general public and especially the young will be considered at the end of this section.

Three broad levels of employment in forestry may be distinguished. *First*, there are the senior *forest officers* and planners as well as some engineers, economists and other specialists who are normally required to have a university degree or equivalent qualification. In the case of the specialists, the question arises whether it is preferable to recruit people with a degree in the subject concerned and to let them obtain a good postgraduate qualification in forestry or to reverse the procedure. Experience suggests that this choice matters less than the selection of the right person. *Secondly,* there is the category of *foresters* embracing the local forest managers and supervisors who work under the university trained forest officers and who require at least some general secondary education as well as a high level of technical and practical training. The *third* category consists of the *forest workers* whose range of required skills varies most. Where the mechanization of forest work has developed farthest as in North America, parts of Western Europe and the USSR, many forest workers are highly trained machine operators whose earnings are of the same order as or even higher than those of their supervisors, while in developing countries most forest workers are unskilled, without any special training and often illiterate.

6.4.1 University level (forest officers)

A forestry degree course at university level generally has three main components: biological, technological and economic. The biological component is concerned with the growing of trees (silviculture) and supporting science subjects such as botany, ecology and soil science. The technological aspect concerns such matters as timber harvesting, wood sciences and forest road construction. The economic component is important because most forest officers are concerned with management, including marketing and long range planning. Major environmental and social considerations enter into all three components. A good university degree course in forestry is thus a sound and widely based education.

In some countries the main emphasis has been on the biological sciences and in others on technology. In many, the economic, environmental and social implications of forestry continue to receive insufficient attention and there is still a tendency for courses to be too discipline orientated and not sufficiently problem orientated. Inter-disciplinary problem solving, incorporating an element of research, is an essential aspect of any forest officer's training; indeed this approach is needed nowadays in almost any responsible position in forestry as well as in other walks of life.

There is a great danger of overloading courses and lengthening them unduly by trying to accommodate all that is thought to be desirable. Most universities realise that an understanding of broad principles combined with a capacity for clear thought may be more useful than too much detailed knowledge of facts which can in any case be found in books and may become obsolete, but they do not always find it easy to reconcile the conflicting claims of lecturers in the diffirent subjects.

Universities themselves will normally be free to decide the contents of degree courses, but government influence may be decisive in determining where they should be given, because it is the government which, with few exceptions, holds the purse strings. Forestry at university level is taught at:

– institutes or forestry faculties located at some distance from a university;
– at universities, as a separate degree course;
– at universities, as a specialization in some broader degree course such as agriculture, land use, biological sciences or engineering.

The isolation of a forestry faculty from other university faculties has obvious disadvantages at a time when the links between forestry and other disciplines become closer and forest officers are required to become more outward looking.

It also renders more difficult any pooling or sharing of teaching resources and other facilities. There may, however, be overriding social and political reasons for such a solution. Thus the Swedish authorities recently moved the Royal College of Forestry from Stockholm to the remote and sparsely populated North of the country as one of several measures to develop a cultural centre in that region.

A specific forestry degree may enable a graduate to become fully effective professionally quicker than the graduate who has taken a more general degree with only some specialization in forestry; the latter on the other hand, may find it easier to adapt to changing conditions, to relate forestry to broader issues and to find employment outside forestry if desired. Placing forestry into a wider context may also lead to some economies in lectures and facilities. More important perhaps than the precise syllabus of a degree course is the standard of teaching.

All around the world there has been a tendency to provide degree courses in forestry (or including forestry) at far more universities than is necessary with the inevitable result that resources are wasted, standards lowered and that only a small proportion of forestry graduates can ever practice their profession. Sometimes there may of course be plausible reasons for the over-provision; for example, in countries where not all inhabitants speak the same language or are of the same race it is understandable that each language group wishes to have its own facilities. Switzerland has, however, demonstrated that a single establishment offering a forestry degree can successfully cater for all its language groups and Luxembourg has demonstrated that a small developed country may find it more satisfactory to send forestry students to universities abroad for their training instead of establishing training facilities of its own. Usually the over-provision is simply due to the reluctance of a particular region to depend on facilities provided elsewhere or to the understandable ambition of a university to provide as wide an option of courses as possible. Governments may have to bow to such pressures but they should at least not be unduly swayed by the argument which is sometimes put forward, namely that the degree course should be tailored to fit the physical conditions of the region where the students will subsequently work. A good university course will enable forest officers to become effective in moist tropical forest, semi desert or in temperate forests; they will of course take a little time to get adjusted to unfamiliar surroundings. The implications for developing countries which have not yet got their own forestry degree facilities are clear: the building up of a cadre of highly qualified forest officers, which must be a top priority because these are the key to the development of forestry, can best be achieved if suitable candidates are sent abroad for training, preferably to a variety of countries, until first class facilities can be created at home.

6.4.2 Technical level (foresters)

Quite different considerations apply to the training at technical level of local managers and supervisors (collectively referred to in this context as foresters). To be effective, these must have an intimate knowledge of local conditions and practices. As their work brings them into close direct contact with the work force and other inhabitants of the area an intimate knowledge of local customs and language are also a great asset. There is thus a very strong case for providing this type of training facility as near to the area of likely future employment as is practicable. This is made easier by the facts that three to five times more foresters are needed than forest-officers and that the teaching staff require less academic specialization and may largely be drawn from practice. It is in fact an advantage if they are. In many countries all over the world training at this level leaves much to be desired. Some countries have sought to achieve the sensible objective of raising the status of foresters by the mistaken method of making the training too theoretical. Foresters then become second rate forest officers instead of experts in their own right. Filling forester posts with university graduates who have failed to obtain forest officer posts produces an analogous result, but is the inevitable consequence of an excessive number of forestry students at universities. Further down the scale, there is a good case for filling the gap between forester and forest worker by giving short courses of a few weeks' duration to selected forest workers. Such courses train workers to become foremen and to give these some mark of authority. Especially where work is scattered, some intermediate rank between the forester and the forest worker is needed. But it is only the exceptional man who can satisfactorily perform a forester's duties with such a rudimentary training.

In the training of foresters the question again arises whether it should be completely separate or whether it should be linked in some way with the training of technicians in related fields: rural engineering (irrigation, water supplies, road construction), agriculture etc. Separate facilities have the advantage that the training syllabus can more easily be tailored to meet forestry needs and be controlled by the forest administration which is usually the main employer. The broader approach has the advantage of preventing forestry trainees from becoming too inward looking: forestry students are in close contact with likely future colleagues in related fields. The broader approach also makes it easier to ensure equivalence of standards between forestry and other qualifications and some savings are achieved through the sharing of facilities and teaching staff by several disciplines.

Wichever approach is adopted, the training should embrace practical forest management (silviculture, harvesting, marketing) appropriate to local conditions as well as elementary planning and office management (organization of work, keeping accounts and records etc.) The mechanical and engineering content will depend on the local mechanization policy. In countries where logging is highly mechanized, there may be a case for having separately trained technicians for the purpose and to concentrate the mechanical training of foresters mainly on the mechanization of silvicultural operations, such as preparation of the ground for planting (scrub clearance, drainage), early thinnings, pest control etc. Under most conditions, however, a forester should be capable of organizing and supervising all harvesting operations. Where, as for example, in parts of Indonesia, there is a conscious policy of retarding mechanization in order to maximize rural employment, the efficient use and maintenance of hand tools should continue to constitute an important element of a forester's training. What a forester should know about engineering matters such as forest road construction or erosion or avalanche control, will again depend on local conditions. There is always a temptation to overload a syllabus, usually at the expense of the practical side of the training which for foresters is all important. The main policy objective of the training of foresters must be to equip them in a course of about two to three years duration to be efficient all round practical forest managers who are respected by their superiors, their workforce and the community in which they live.

6.4.3 Operational level (forest workers)

The training of forest workers must be even more closely linked to local factors, than that of foresters and must take into account their level of education, their practical experience of forestry and use of machines, whether or not they are engaged in forestry full time or part time, and so on.

The efficiency of forest workers as of any other workers depends much on job satisfaction, and this depends at least in part on proper training and the earning power that goes with it. There are, of course, also other factors. In some of the poorest countries for example, training must be accompanied by remedying nutritional deficiencies.

Where workers are unskilled and perhaps illiterate, training must provide immediate tangible benefits, proceeded only by small and easily understood, well planned steps from their accustomed work habits, and the instructors must be people they know and trust, preferably from their own community or tribe, who speak their mother tongue. If forestry training can be put into the context of rural development as a whole so much the better.

Even in developed countries there are still many relatively unskilled forest workers but the problems of training are somewhat different because of the higher educational standards on the one hand and the more sophisticated methods of working on the other. The main beneficiaries of training are the workers themselves whose work becomes more skilled, more productive and better paid; but also forest owners and industries benefit directly from a more productive, better paid and contented labour force. State forest enterprises will necessarily have to make arrangements for the training of their own workers and it may be advisable for these facilities to be made available to the private sector since only large organizations are normally in a position to run their own training schemes.

The operators of very large and sophisticated machines are usually recruited from the most highly skilled workers who already have mechanical experience; the further special training required to operate such machines, which is beyond the scope of ordinary forestry training, is sometimes provided by the machine manufacturers.

The training of part time workers, such as farmers with small wood lots, presents some special problems: the time such workers can spare for training in forestry may be limited and restricted to certain seasons when they are not otherwise engaged; they may not be able to afford to pay for the training or even to forego earnings while it takes place; they are even less likely to be able to organize such training among themselves. Employers for their part find it less attractive to arrange and pay for the training of part time than of full time workers because the benefit to the employers is more limited. The training of part time forest workers, including small farmers or others who are self employed, is of public interest because without such training, a nation's forest resources may not be fully utilized. Especially in the case of small wood lots, skilled part time workers are in the best position to bring wood onto the market economically.

The degree to which the state forest authority should get involved in forest worker training is a matter of judgement, but the authority must accept responsibility for ensuring, in consultation with all interested parties including the trade unions or other organizations representing forest workers, that adequate training facilities and standards are maintained, with such technical and financial assistance from the state as may be required. A particularly important duty is to ensure that all training includes adequate precautions against ill health and accidents at work and the correct action to be taken if an accident or injury does occur.

6.4.4 Refresher courses

In forestry as in other walks of life, the pace of technological progress has accelerated and will continue to do so. Refresher courses at all levels are therefore becoming more and more essential. Every forest officer and every forester should be able to spend a few weeks every three to five years on such a course. Experience suggests that such courses achieve the best results if each is orientated towards some specific aspect of forestry e.g. plantation management, erosion control, or forestry for rural community development. The candidates for each course can then be selected from among those who are particularly concerned and interested. Since such courses must necessarily be short if the participants' normal work is not be unduly interrupted, more general courses tend to be too superficial to achieve much. Refresher courses may be organized by the universities and forester and forest worker training establishments where the initial courses are given, or by forest administrations; in either case refresher courses gain greatly in value if outside experts with first hand practical experience of recent developments are brought in to assist.

6.4.5 Courses to obtain additional qualifications

Quite distinct from refresher courses are the usually much longer courses leading to additional qualifications. An increasing number of forestry graduates now aspire to some additional formal qualification such as a doctorate. In most cases these graduates depend on government or some national or international organization or institution be bear the cost. There is a very good case for enabling graduates to obtain such additional qualifications if they are to occupy specialist posts, e.g. in research or as economists. Where this situation does not apply, it may sometimes be preferable to give graduates the opportunity of a travelling fellowship on which they may learn at first hand how certain problems are solved elsewhere. There is of course the danger that, unless precautions are taken, travelling fellowships may simply result in tourism. Normally the best time for additional courses or travel fellowships is when a candidate has had at least a few years experience in a responsible post but is still sufficiently young to absorb new ideas and put them into practice.

Similar considerations *mutatis mutandis* apply to the further training of foresters. But here there is also the additional question of enabling those who appear particularly suitable to qualify for forest officers' posts. Foresters who are sufficiently young and have the necessary educational qualifications are best given the opportunity of going to a university. Those without such qualifications should be given some other opportunity to widen their horizon such as a travelling fellowship before they are promoted. Some forest officers who have received their promotion in this way have been excellent, but where this door has been opened too far, there has been a tendency for the standard of forest officers to decline.

Forest workers are often promoted to junior supervisory posts without any formal additional training, but where a young worker displays particular promise he deserves to be given every opportunity to develop his full potential by the attendance of courses leading to appropriate qualifications. In this way some forest workers have become good foresters and a few even good forest officers.

6.4.6 Informing the public

Ignorance and apathy are among the foresters' main enemies, because only people who have some understanding of forestry have an incentive to play their part in the prevention of forest destruction (e.g. by precautions against fire) and in the support of positive forestry measures.

The imparting of knowledge and promotion of understanding start with the young. It is as part of nature studies – an essential although sometime neglected subject at school – that the role of forests should be explained and demonstrated, preferably not only in the classroom but also in the woods themselves. Visits to woods are particularly important for children in large towns, who have little personal experience of the countryside.

The forest authority will rarely have much direct influence over what is taught at schools, but it can and should help by making available suitable teaching and demonstration material and by encouraging and providing facilities for visits to woodlands, conducted if possible by forestry staff. Experience suggests that explanations by people, who are, or have recently been 'on the job' are particularly appreciated by the young. In some countries retired foresters are said to play a very valuable role in conducting school excursions.

Among grown ups the opening up of forests for recreation has probably done more than anything else to arouse an interest in and understanding of forestry, especially where the opening up has been accompanied by the provision of information centres, nature walks with explanatory notices and amenities such as car parks and picnic sites. Good television programmes about forestry, well illustrated publications, lectures, the organization of tree planting days are among other methods that have been proved helpful.

To foster public relations and to spread knowledge about forestry among the public should be clearly defined aspects of policy. The implementation of this policy should be the specific responsibility of one of the forest authority's senior officers. Whether this should be a full time or a part time assignment will depend on circumstances and so will the question whether there should be any permanent staff to assist this officer. Professional public relations officers are generally best at dealing with the media – radio, television, press – and also in the planning and coordination of the various activities involved, but experience suggests that forest officers and foresters may often be forestry's most effective ambassadors to the general public.

6.5 Research

The important role of research in forestry is now generally recognized but the support for it often falls short of what is needed. The main policy problems are concerned with the allocation of resources to research, decisions on research priorities and the transfer of research results into practice. To put the policy aspects into perspective we must first look at the history, scope and organization of forest research.

6.5.1 History

When forest research began in Europe in the 19th century, it was confined mainly to questions of silviculture and management: how to secure a regular and if possible rising yield, how to protect the forest against pests and fire, and how to enable forests to reduce or prevent erosion. The emphasis was on improving the natural forest. A new era of research opened when foresters started to do what farmers have done for many centuries, namely introduce exotic tree species and breed more efficient varieties to achieve their objectives. A further new dimension was added to forest research by environmental and social forestry and the recognition that timber is only part of the renewable biomass produced by trees. The mechanization of forest operations and especially of the harvest and transport of timber also introduced additional research disciplines. These trends have immensely widened the scope of forest research, shifted a significant proportion of it from the forest to the laboratory and brought it into closer contact with many other fields of research, ranging from biochemistry and genetics to ergonomics, engineering, and landscape design. There are now also forestry components in many other fields of research.

6.5.2 Scope

The main areas of forest research are as follows:

1 biological: silviculture and tree breeding including choice of tree species, selection of origins and individual trees for the breeding of improved strains and clones; fertilization; studies of growth and yield; protection against abiotic and biotic damage; effect of silviculture on wood properties (see also 6 below);
2 environmental: protection of the environment including soil and water conservation, wildlife management etc.;
3 technological: mechanization of forest operations including the harvesting and transport of timber, forest road construction etc.;
4 social: safety and health of forest workers; the role of forestry in promoting the well being of rural populations; problems connected with recreational use of forests etc.;
5 economic: the application of economics to forest management and the marketing of forest produce;
6 study of wood properties as a guide to utilization on the one hand and silviculture on the other.

Some of the above problems cannot be solved by research alone. Administrative decisions, common sense, practical experience based on trial and error also play a part especially in the solution of the social problems, but even here progress will be retarded without adequate research.

Most forest *products* research is concerned with wood, the main emphasis being on:

1 mechanical processing as in sawmilling, particle board and plywood manufacture;
2 conversion to fibre as in the manufacture of fibre-board, pulp and paper;
3 use as a chemical feedstock for the manufacture of a wide range of chemicals (see Chapter 3.1);
4 use as a source of energy, either by direct combustion or via transformation into fuels such as methanol.

There are, however, also a very large number of at any rate locally important forest products other than wood, which call for research: bamboo, ratan, resins, gums, numerous medicinal forest plants (especially in the tropics), fruit, fungi, forage for cattle, to name but a few.

In many countries responsibility at government level for forest and forest products research rests with different ministries, because the research links between forest products and other materials are considered to be stronger than the links between forestry and forest products, which are perhaps more evident to foresters than to others. Even where, as in the USA, the Federal Forest Service has a major responsibility for forest products research, it is usual for this research to be conducted at separate research centres. A notable exception is the Federal Forest and Forest Products Research Institute in the Federal Republic of Germany (Bundesforschungsanstalt fur Forst-und Holzwirtschaft, near Hamburg), which combines both. The relative advantages and disadvantages of linking forest and forest products research seem to depend very much on local considerations such as the source of funds. In what follows, forest research only will be considered, because in most countries those responsible for forest policy have no direct responsibility for forest products research.

6.5.3 Organization

Forest research is carried out mainly by government forest research institutes and by the forestry faculties of universities but many research organizations which are not primarily concerned with forestry undertake research of relevance to forestry. For example institutes specializing in any field of biological research will naturally wish to include in their activities species and vegatation types of forestry importance. These contributions to forest research can be very valuable, but it is only at institutes or university departments dedicated to forestry as a whole that the relationships between the different facets of forest research can be dealt with satisfactorily. That is why the major research effort in forestry should be at such centres.

A point that tends to be overlooked nowadays, both by forest administrations and research organizations, is the great contribution to the advancement of knowledge and forestry practice, which has been made and can still be made by forest officers who are not part of any research organization, but manage forests. Field officers have developed many successful silvicultural methods all over the world; especially in the moist tropical forests they have added much to our knowledge of the immense number of species and their ecological status. The contribution of field officers will be most fruitful if it is encouraged from above, if the personal links between research officers and field officers are close and if the latter are left long enough in one region to get to know the forests intimately. It is the opportunity for continued observation over a long period which enables a forest manager to make discoveries which might elude controlled experimentation. His most effective contribution is usually in forests about which least is known, but even in countries with a long forestry tradition, his contribution need not be negligible.

Government forest research institutes are usually concerned mainly with *applied* research, in support of specific policy decisions (eg to afforest certain types of land) or to solve specific management problems (eg to reduce the cost of harvesting small trees). *Universities* are mainly concerned with more *fundamental* research of their own choosing. Research is an essential adjunct to university teaching and since academic freedom lies at the very root of university life, it follows that this freedom must apply to research as well as to teaching. That of course does not preclude universities from voluntarily accepting research assignments suggested and perhaps paid for by government or other bodies.

To oversimplify: applied research is mainly concerned with answering the practical question *what* should be done? while fundamental research is mainly concerned with answering the more theoretical question *why*? While applied research usually leads to quicker practical results, major new developments are often only made possible by fundamentel research. The distinction between applied and fundamental research is not absolute but rather of emphasis and the two are complementary. For example applied research may show that a given silvicultural treatment influences wood quality in a certain way. Fundamental research may explain the reason why and thus lead to a further inprovement in silviculture.

Of course, a good scientist engaged on applied research will look beyond the achievement of his immediate objective and in a good organization he will be encouraged to do so. Conversely a good scientist engaged on fundamental research will be attracted to projects which offer some prospect of results which are likely to benefit practice, at least in the long term.

Governments cannot-or at least should not-influence unduly research done at universities or other independent research organizations. They must therefore have their own research capability in forestry if their own research priorities are to be met. A very satisfactory arrangement which exists in some countries is for the forest authority to have a research organization under its direct control. This arrangement facilitates – but cannot guarantee – the correct allocation of research priorities, close contacts and even transfers between research and field staff, transfer of research results to forestry practice and the feed back of experience and problems from the field; the administrative difficulties which may arise over matters such as the provision of sites for experiments in state forests, are also minimized. Some countries, however, have preferred other arrangements. In France, for example, the government forest research organization was transferred some years ago from the forest authority to the National Institute for Agricultural Research (Institut National de la Recherche Agronomique (INRA) which, in common with the forest authority, comes under the Ministry of Agriculture. As forest research and agricultural research have some elements in common, organizational links may achieve some economy of effort and prevent forest research from becoming too isolated and inward looking. On the other hand, there are also certain risks. Apart from the direct link with the forest administration being lost, forestry may receive insufficient attention in a research organization in which it is only a minor component; moreover, apparent similarities of research problems in forestry and agriculture may obscure fundamental differences. For example, both are concerned with plant breeding, but much of the research appropriate to annual crops may be quite inapplicable to trees – a point which even experts sometimes do not fully recognize. On the other hand, geographic arguments against having forest and agricultural research institutions in the same place are sometimes exaggerated. As forest experiments are usually spread over a wide area, the location of the research headquarters at or near an agricultural research centre is likely to cause inconvenience only if most of the forests happen to be at the opposite end of the country.

The arguments for and against organizational links between forest research and agricultural research would, *mutatis mutandis*, also apply to similar links with other research institutes with some topics in common, for example, institutes dedicated to environmental research.

6.5.4 Policy priorities

Research policy must be forward looking. In addition to trying to solve today's problems an attempt must be made to anticipate tomorrow's. Forest research must in the first place be geared to the achievement of the goals of forest policy but it should also seek to contribute to the general advancement of knowledge and the quality of life.

To achieve these objectives forest research must be adequately endowed with personnel and funds, it must be well organized and directed and there must be adequate links with research in related fields. It is sometimes not appreciated that the best research workers are not necessarily the best directors and organizers of research.

Personnel. Orginally, forest research was conducted almost exclusively by forestry graduates; but some of the more advanced and specialized problems of modern times can only be tackled by interdisciplinary teams of scientists. Even so, only a strong core of forestry graduates, preferably with practical as well as research experience, can ensure that the research remains practice orientated. Because these forestry graduates can easily be transferred from research to practice and *vice versa* they facilitate the transfer of results to practice; they can also contribute to a better understanding of management problems within research organizations. Too frequent cross postings are of course disruptive.

Forest research requires not only university graduates but also good foresters and other technicians for the lay out, measurement and maintenance of field experiments and for various kinds of laboratory work as well. Foresters and technicians are more specifically trained for these practical tasks than graduates and cost less. The use of field staff instead of research staff for the establishment of field experiments may save money and strengthen the links between research and practice, but the field staff usually lack the time and training to meet the necessary standards.

The funding of forest research raises a number of issues. First and foremost, erratic changes must be avoided. Most projects can only give results after a number of years; if funds are cut off before completion, much of the money already spent will be wasted. Most forest research is funded by governments, but universities, independent research foundations and institutes, forestry and international organizations also contribute funds.

Governments generally channel the bulk of their expenditure on forest research to their own research establishments for the purpose of achieving specific forest policy objectives, but they usually also contribute financially to research that is done elsewhere, serves wider objectives including the general advancement of science and is at the recipient's discretion. Here we are concerned primarily with research as an instrument of forest policy. The governments of rich countries can afford to spend more than those of poor countries, both in absolute terms and in relation to total forestry expenditure, and they usually are also in a much better position to undertake research. They have more qualified scientists, better laboratories and a more advanced institutional infrastructure. They thus have a particular advantage in fundamental research which generally requires highly specialized scientists as well as elaborate and expensive equipment. The advantage over developing countries is much less in respect of empirical applied research, much of which can be conducted with relatively inexpensive equipment by officers with a good forestry or other general scientific background. Developing countries are therefore well advised to concentrate their limited resources on this type of research. Their forest managers must also bear in mind that it may sometimes be preferable to act on imperfect knowledge, at the risk of making mistakes, instead of awaiting the results of research by which time action may no longer be possible, because the forest has been destroyed.

A common defect of funding arrangements in developing countries, is the allocation of too large a proportion of scarce funds to salaries and too small a proportion for the equipment, transport and everything else that is needed for the conduct of experiments in the field, the only place where most applied research and even some fundamental research can be carried out.

Research programmes are an inevitable compromise between what is thought desirable and what can be achieved with the available resources. The identification of priorities is thus a key issue. An attempt must be made to find the right balance between short term and long term objectives, between applied and fundamental research, between the major research disciplines. In deciding these issues it is important to bear in mind the following points:

- a research capability in the more specialized areas of research cannot be built up at short notice nor can specialists in one discipline (e.g. tree breeding) readily be transferred to another (e.g. entomology or harvesting technology;
- there must be a planned phasing out of research that is no longer relevant (e.g. on species no longer used) or that has fulfilled its main purpose;
- not all duplication of effort is wasteful because the approach adopted by one

research centre may prove to be more effective than that followed by another, but unnecessary duplication (of which there is plenty!) should be avoided;
- before embarking on research it is useful to see what others have done and to build on it; there is no need to 're-invent the wheel'.
- the objectives of a research organization may not coincide with those of individual research workers who may be more interested in obtaining academic distinction and opportunities for foreign travel than in solving problems of importance to practice; that is one reason why, especially in developing countries, there tends to be an unwarranted bias towards theoretical research of little practical value.

For *reviewing research programmes* it is advisable to seek the advice of scientists from outside the research organization concerned. This may be done on an *ad hoc* basis or by appointing a permanent advisory committee. In the case of government forest research institutes it is also highly desirable for the forest authority to exercise some direct and effective control over the broad research priorities in order to ensure that they are in line with forest policy priorities; this control should, however not lead to interference with the research itself.

There is also a strong case for a periodic general review of a country's forest research as a whole in order to ensure that it may contribute effectively to forest development as well as to the general advancement of science.

The application of research results frequently leaves much to be desired. The results are generally of interest
(i)   to forest managers who apply the results in practice;
(ii) to other research workers as a starting point for further research.
Forest managers do not need detailed accounts of the research work itself; in fact they rarely have either the time, inclination or special knowledge for that sort of reading. What they need are concise guides to practice based on the research results and supplemented where practicable by field demonstrations, lectures and personal discussions with the research workers concerned. Many scientists do not find it easy to write concise guides to practice and some research organizations have appointed special staff for the purpose. There must be no undue delay in this transfer of knowledge; preliminary indications of provisional results are often useful. The flow of information must of course not be only in one direction. A feed back of observations from the field to research can be a valuable stimulus to further research.

Research workers who wish to follow up a particular line of research need a detailed scientific account, but there is a much wider circle of research workers who will merely need a summary.

6.5.5 International co-operation

The exchange of information and co-operation between forestry research institutes is greatly facilitated both nationally and internationally by the fact that, as a rule, no financial advantage is to be gained by keeping research results secret. Recent exceptions are the patented results of genetic engineering obtained by some private research organizations; such exceptions may spread but are unlikely to become of major importance. This being so, virtually all governments encourage international co-operation in forest research under the auspieces of the International Union of Forest Research Organizations (IUFRO) and other international organizations whose activities are described in Chapter 7. In fundamental research no countries, except possibly the super powers, can afford to maintain a really efficient capacity in all aspects of relevance to forestry and even the super powers will benefit from co-operation and a partition of labour.

The willingness to collaborate is of particular significance to developing countries whose research capability is limited. Research organizations in a number of developed countries are willing, indeed eager, not only to transfer research results but also actively to co-operate in research projects with developing countries, either internationally or bilaterally; scientifically, such partnerships can be as rewarding to the donor as to the recipient organizations. A form of co-operation which is now widely thought to be very effective is the 'twinning' on a continuing basis, between one institute in a developed country and another in a developing country. This is a view also supported by IUFRO. To get the maximum benefit from research aid or partnerships, developing countries must be clear about their research priorities, because these may not always coincide with those of their partners. In this context, it is encouraging that the question of forest research priorities for developing countries was discussed at the IUFRO World Congress at Kyoto in Japan in 1981, on the basis of a paper prepared under the auspices of the World Bank and FAO. The Congress endorsed the findings of the paper which identified three main topics that are deserving of special attention in the coming decade. These are:

- research related to the contribution of forestry to rural development, i.e. the contribution of forestry to meeting the production, income and environment needs of rural people;
- research related to energy production and use, into ways and means of increasing the productivity of trees to produce the maximum biomass and energy yield, and into conserving wood resources by more efficient use of wood for energy;
- research related to more effective conservation and management of tropical forest eco-systems with special reference to protection of the environment.

6.6 The government forest services

There are two main functions which a government *forest service* (sometimes also referred to as forest administration or forest department) may be called upon to perform:

- as *state forest enterprise* to manage the state forests and, in some cases, also other publicly owned forests;
- as *forest authority* to advise government on forest policy and to administer its implementation. Forest authority responsibilities thus generally include the enforcement of forest legislation, the administration of aid schemes for private woodland, relations with various government departments and non-governmental organizations as well as with other countries and international organizations.

In some countries, for example Great Britain, a single forest service performs both the forest authority and the state forest enterprise functions. In other countries such as Austria, France and Sweden there is a separate forest service for each.

The terms 'state forest enterprise' and 'forest authority' will be used whenever there is a need to differentiate between these functions and the more general term 'forest service' when a statement applies to either or to both. Ministerial responsibility, and the structure of forest services are the main policy issues. Both are so bound up with a country's machinery of government that the scope for general advice is limited.

6.6.1 Ministerial responsibility

A separate minister for forestry or for forestry and forest industries combined is unfortunately, a rarity. The more usual arrangement is for the minister of agriculture or some other minister to have forestry as part of a larger portfolio. This divergency in practice reflects the difficulties of finding a satisfactory arrangement.

From the forestry point of view there are obvious advantages in having a separate cabinet minister to defend the interests of forestry at the highest government level, but forestry usually plays too small a part in national life to warrant separate representation in the cabinet. In that case there should at least be a junior minister with forestry as his sole responsibility.

The common practice of placing forestry under the minister of agriculture has both advantages and disadvantages. The arrangement is logical because farming and forestry are complementary in the use of land resources and to some extent also as sources of employment. To quote just a few examples: there are farmers who own woodlands and forest workers who have small agricultural holdings; woodlands benefit neighbouring farmland by providing shelter; foresters and farmers have common interests in rural development. The main disadvantage of the arrangement is that farming tends to steal the limelight while forestry suffers from benign neglect and loses out whenever there is a conflict of interests. These risks are less serious where many farmers also own woodlands.

Other ministries into which forest services have been placed include; Environment, Finance, Home Affairs, Land Use, Renewable Natural Resources. Ministries such as those of Environment or Finance are likely to emphasize particular aspects of forestry; a large ministry such as Home Affairs may be suitable in a small country where forestry is important, such as Luxembourg, but might neglect forestry in a large country. Where there are ministries either for land use or renewable natural resources, these offer perhaps the most suitable home for forestry, especially if there is a separate ministry for Agriculture.

A very bad system is for the forest authority to be responsible to different political masters in respect of the productive and service functions of forestry. This makes it extremely difficult, if not impossible, to develop and implement a coherent forest policy with a sensible balance of emphasis between the various functions. These objections apply less to a territorial division of responsibilities as in Britain where the Forestry Commission is responsible to separate ministers in England, Scotland and Wales.

6.6.2 Forest authority and state forest enterprise

The *forest authority* advises government on forestry matters and implements forest policy decisions taken by government. Given the multiple objectives and functions of forestry, the government should seek and take advice on policy also from various other sources both governmental and non-governmental as will be discussed in more detail in Chapter 8. A policy, once decided upon can, however, only be implemented efficiently by a single authority, namely the forest authority. Otherwise there is bound to be inefficiency and friction just as there will be if the forest authority is responsible to different ministers or ministries in respect of the different functions.

While the allocation of responsibilities is clear in principle, the application of these principles is less so. Wildlife management is a case in point because wild living animals and plants do not respect forest boundaries. In many countries the forest authority is responsible for the implementation of hunting and other wildlife legislation both inside and outside forests. There is some logic in this arrangement, but it has not always worked well in practice when forest authorities have been too weak to resist the political pressures from strong hunting or conservation lobbies.

Inland fisheries are sometimes also the responsibility of the forest authority as for example in Belgium and France. This and other similar additions to the responsibility of the forest authority may be very appropriate. If they go so far that forestry becomes a mere component in a much wider service, for example, a service responsible for all nature conservation in a country there is the risk that forestry may receive too little attention.

As already mentioned, the responsibility for managing the *state forests* rests in some countries with the forest authority, while in others the responsibility is vested in a separate state forest enterprise. Even where the forest authority combines both functions, these may be separated administratively by confining the responsibilities of some forest officers to either one or the other function.

An argument in favour of separating the functions is that the state forest enterprise might be in competition with private woodland owners for example over the marketing of timber or over the purchase of land for afforestation; and that the forest authority might in that case consciously or unconsciously favour the state forests if it is responsible for them. In countries where the actions of government can be criticized freely, there is little risk of private interests being harmed in this way, and actual examples of harm are difficult to find. It may, however, be true that a separation of functions reduces the risk of either the one or the other being neglected, since some forest officers prefer spending their time managing forests while others prefer the advisory, supervisory and inspectoral tasks of the forest authority. The task of managing state forests commercially may be easier if they are separate or failing that, at least have separate budgets and financial control.

The combination of the two functions, however, also brings some definite advantages. Especially in small countries or countries with limited or widely scattered forest resources, there will be a saving in personnel and other administrative costs. Forest officers with management responsibilities of their own may also be in a better position to understand the problems of private woodland owners, to offer practical advice and guidance and to help for example with fire prevention and control or by reducing or even halting normal fellings after a storm catastrophe.

Most countries appear to be well satisfied with whatever arrangement they have; those who are not, would do well to study the experiences of others before embarking on any change. In small countries and in countries with recently established forest services the balance of advantage is likely to lie with a single organization for both functions, especially if forest and financial resources are limited and there is as yet a scarcity of highly qualified and experienced forest officers.

6.6.3 Structures

Virtually all forest services have a headquarters organization at the centre and one or more levels of regional and local offices to cover the country as a whole.

The forest authority is usually treated like any other government department whose head is responsible to an even more senior civil servant or, in some instances a minister. A similar arrangement may apply to a separate state forest enterprise, but in some countries the direction of the state forest enterprise and more rarely of the forest authority is vested in a board of management along the lines of a public corporation or a private company. This type of structure facilitates management along commercial lines; and a forest authority with this structure is likely to have more independence and scope for initiative than it would have as part of a ministry. In practice, the influence of the Board will depend largely on the status and influence of the people appointed by government to serve on it.

The number of steps in the territorial sub-division of a forest service depends on physical factors such as the size of a country and the nature and extent of its forest resources as well as on historical and constitutional factors such as the degree of autonomy enjoyed by provinces or other relevant constituent parts. If forestry conforms to the territorial sub-divisions of general administration, contacts and collaboration with other government departments are made easier. These advantages may however be outweighed by other considerations such as the desirability of reducing the number of steps in the territorial hierarchy. In Great Britain, for example, the Forestry Commission (which is both forest authority and the state forest enterprise) abolished its separate offices for England, Scotland and Wales during the 1960's. This shortening of the chain of command led to some saving of staff and to an increase in efficiency without causing significant administrative of political problems.

The distribution of responsibilities between headquarters, regions and districts will depend largely on a country's constitutional set up. In countries with a highly centralized general administration, the forest authority function and the management of the state forests also tend to be highly centralized.

In countries with a federal set up, on the other hand, the main weight of responsibility in forestry is, as one might expect, at regional level and the central headquarters are concerned only with general policy, legislation, external relations and, in some instances, with research. The larger the country and the greater the geographic and climatic differences within it, the stronger is the case for decentralization. Experience suggests that forest management should be more decentralized than research and other specialist activities. Whatever the organizational set up, there is a great advantage in securing the mobility of staff between regions; otherwise forest officers may become too parochial and inflexible in outlook. The matching of candidates and posts also becomes difficult without an adequate degree of staff mobility; this applies especially to senior appointments which are in any case relatively small in number.

The headquarters of the *forest authority* should be within easy reach of the office of the responsible minister and of other centres of political power; forestry's influence is likely to be proportionate to that proximity. Offering a forest authority an attractive headquarters in a beautiful rural setting might be the most effective way of preparing it for a relatively painless death. The same may not apply to the *state forest enterprise* whose management task may in fact be easier to accomplish at some distance from political and other pressures.

At regional and local levels the links between forest services and other services such as those for agriculture, are most easily maintained if the offices are in physical proximity to each other, preferably in the same building. Where the forest authority function is separated organizationally from the management of the state forests, physical proximity of the offices of the two organizations is also likely to be an advantage.

Some countries have attempted to strengthen interdepartmental links at regional level by incorporating forest offices and other offices concerned with land resources in a regional set up under a senior official who is directly responsible to central government. This may facilitate interdepartmental co-ordination in the region, but usually at the expense of forestry which carries less weight than agriculture. Moreover, the system easily obscures lines of command and the senior regional forest officer may get conflicting directives from his head office and from the head of the regional set up.

6.6.4 Finance

Forest authority work is almost invariably financed and accounted for year by year in the same way as other government expenditure. The state forests are either dealt with in the same way or some attempt is made at accounting along the lines of a commercial enterprise. State forest enterprises understandably argue for a system which will give them the maximum independence and assurance of continuity of funds. They argue that freedom from excessive bureaucratic control is essential for efficient management and as an incentive to good performance; they also point to the waste caused by sudden fluctuations in available funds: For example, money spent on establishing a plantation may be wasted if no funds are subsequently made available for its maintenance and protection against fire. Ministries of Finance, on the other hand, like to keep a tight hold on purse strings and point to the undesirability of exempting forestry from budget cuts in an economic recession. In a few countries where annual forest revenue exceeds expenditure, forest services have succeeded in obtaining an arrangement whereby part of the surplus may be retained for re-investment in forestry and as a reserve to be drawn on in lean years, but such arrangements tend to be withdrawn by government if there is a serious crisis. Where annual expenditure exceeds revenue, the case for long term funding is even less popular with governments and the need for a clear separation between forest authority finance and state forest enterprise finance correspondingly greater.

The fundamental point to bear in mind in deciding on budgets for forestry is that forests are not only expected to produce timber, but to provide also the benefits and services described in Chapter 4, which are important but are difficult or impossible to evaluate in money terms. What, for example, is the value of preventing the depopulation of a rural area of the survival of endangered species of wild life?

6.7 Personnel

6.7.1 Professional, technical and administrative personnel

Manpower planning presents few problems in long established forest services whose activities have continued at more or less the same level over a considerable period, and where the distribution of staff is more or less normal in terms of age and seniority. It is then relatively easy to predict for a number of years ahead the number of posts that will fall vacant through retirement or other causes, the promotions that can be made and the need for recruiting new staff. In recently established and expanding forest services, staff planning is much more difficult and its importance is often not fully appreciated. Too little account is taken of the longer term consequences of ad hoc decisions. Understandably, an effort is made to build up a cadre of qualified staff as quickly as possible, but sometimes it may be preferable to leave posts unfilled rather than to accept applicants with lower qualifications than those appropriate for a post. Each case must be considered on its merits, but unless there is an insistance on high standards, the efficiency and standing of a forest service is bound to suffer.

The second question, which is to some extent linked to the first, relates to the rate of expansion. If expansion is too slow, the momentum is lost; on the other hand, rapid expansion, even where suitable candidates are available, may cause organizational problems in the short term and personnel problems in the longer term, because a time of rapid expansion is bound to be followed by a more static period, during which most posts will be occupied by people who are still relatively young. There will thus be few retirements and recruitment must drop drastically. If, meanwhile, training facilities have been geared to the recruitment levels of the expansion phase, there will be far more qualified candidates than jobs. Moreover, as even most officers in senior posts will be relatively young, promotion prospects for officers in junior posts will be negligible and morale may suffer. Rapid expansion can also lead to a temporary reduction in efficiency because the experienced officers have to devote much of their time to helping the newcomers. A third question concerns the relationship between manpower planning and financial planning. It is unrealistic to plan an expansion of staff without considering the longer term financial implications, and the likely availability of funds;

unfortunately, this is frequently ignored.

Other questions of personnel policy include: the proportion of university trained forest officers to foresters with a technical training; the extent to which specialist posts, e.g. in economics or ecology, are filled with forest officers or with graduates in the respective specialization; promotion procedures; mobility within the service; and in service training. The answers to these and similar questions must, to a large extent, depend on local circumstances. Nevertheless, there are a few general points to note.

First, controversy sometimes arises over the question whether or not the post of head of forest service and other senior posts should be reserved for candidates with a forestry degree or equivalent professional qualification. It is widely accepted that the more senior the post, the less important is detailed technical knowledge, and the more important are administrative ability and other attributes of good leadership; but among two candidates of similar ability, the one with a relevant professional background is likely to be the more suitable and it is a sad reflection on personnel policy and career development in a forest service if there are no professional officers who are more suitable for top posts than others without such a qualification. Moreover, morale may suffer and high fliers may be put off a career in forestry if they see the principle applied that 'experts should be on tap and not on top.'

Party politics should not enter into the filling of senior posts. A long term enterprise such as forestry can only prosper if there is a continuity of policy and of senior personnel when there is a change of government. Top posts should be filled by capable people who are not too old and who have the prospect of remaining in post for sufficiently long to make an impact but not long enough to 'run out of steam'.

A widespread weakness in forest services is the lack of foresters with a sound, *practical* training to back up the university trained forest officers; such foresters are indispensable for organizing and supervising the work in the forest. Where this weakness exists, the creation of suitable training facilities should receive high priority (see Section 6.4).

Another general question concerns specialization, which is inevitable if technical progress is to be maintained, but which may cause difficulties if taken too far. The holders of posts in logging, economics, ecology, etc. may then no longer be eligible for promotion in line management. This may lead to problems of career structure and may cause the experts concerned no longer to see the forest through their own particular trees; these difficulties can be greatly reduced by the encouragement of interdisciplinary teamwork which is in any case often the best way of solving problems.

The detailed decisions on all these matters fall into the realm of administration, but it is an important and, indeed, vital issue of policy to provide for the administrative machinery which will ensure a high standard of man power planning and personnel management. Forests are unlikely to be well managed unless the men and women entrusted with the task, are themselves properly cared for.

6.7.2 Forest workers

Many of the considerations which apply to professional, technical and administrative staff also apply to forest workers, but the latter also raise some specific policy issues.

In developed countries the role of the forest worker has changed enormously within the last generation. While some forest operations, especially logging in mountains terrain has always been recognized as requiring a very high degree of skill, most forest work was relatively unskilled manual work and poorly rewarded. Mechanization and the introduction of sophisticated manual operations (eg. application of herbicides) have brought about a drastic change. Nearly all forest workers are now expected to be skilled and versatile. When they operate machines in the forest they must cope with breakdowns unlike workers in a factory who are much closer to back up services. The three main policy issues connected with these developments are:

- training facilities and programmes which have already been considered in Chapter 6.4 (education and training)
- the impact on employment
- safety and health

The impact on employment is highlighted by the fact that in most developed countries the number of forest workers was halved after 1960 over a ten to fifteen year period and is still continuing to sink. If human hardship is to be minimized, careful planning is needed in the way jobs are phased out and new developments are introduced. Opportunities for early retirement and part time work, limitations on recruitment, expansion of forestry programmes and the degree and pace of mechanization must be considered. While in some of the major forest countries in the world such as Canada, Sweden, USSR the general trend has been towards the maximum mechanization in the shortest possible time, other wood rich countries such as Finland and Norway have pioneered the concept of *sophisticated intermediate technology* based on the use of smaller machines. In countries with more scattered forest resources as in the Member States of the EEC, the scope for the economic deployment of very heavy harvesting machinery is in any case limited, but even in the US and Sweden intermediate technology continues to be viable under certain circumstances. In both these countries a large proportion of timber, especially from farm forests, continues to be felled by chain saw and extracted by farm tractors with special forestry attachments. To some extent the degree of mechanization is determined by economics. Heavy machines pay in the case of large scale operations, but a farmer finds it more profitable to log his small wood lot using as far as possible the machines he already has. There is now a trend to favour intermediate technology both on environmental and social grounds: the environmental objection to large harvesting machines is that they can only operate economically if large areas are clear-felled at one time; the social objection is that where the opportunities for rural employment are limited excessive mechanization further reduces these opportunities. On the other hand, the procurement of higher standards of living for the work force are necessarily linked to increases in productivity which in turn depend on improved technology.

To provide for the *safety and health* of his work force is the duty of every employer and it is the duty of the government to see that this is done. Forest work, especially logging, has always been among the more hazardous occupations and this has not changed with mechanization although the nature of the risks may be somewhat different. The dangers to health have, if anything, increased (e.g. 'white fingers' and damage to hearing through prolonged use of chainsaw). Proper training is the best precaution against both accidents and ill health, but training must be backed by rules (e.g. wearing of hard hats, earplugs) and measures to enforce them.

In most developing countries a very large proportion of forest workers is still unskilled and in some cases illiterate. Under these conditions clear policies on the degree and pace of technological innovation are indispensible. At one extreme the rapid introduction of advanced logging equipment has sometimes proved a disaster for a number of reasons: trained personnel to operate and maintain it may not be available and spare parts which have to be imported may be difficult to obtain because of currency difficulties or bureaucratic bungling; but there are even more basic objections: workers who are displaced by machines may have no other source of income and the purchase of machines requires foreign currency of which most developing countries are very short.

At the other extreme the mistake is made of continuing the use of hand tools without the slightest attention being paid to their proper maintenance. The effort required to fell a tree or do any other job is thus needlessly increased. Proper tool maintenance would take some of the needless sweat out of forest work and would also reduce waste.

To develop and implement a sensible policy between these extremes is a main task for forest authorities who will have to bear in mind that somewhat different considerations may apply in different parts of a country and that what is most suitable in state forests may not be the best elsewhere.

In developing countries it is also usually very important to link forestry employment to social measures: in the first place it is necessary to ensure the provision of adequate nutritional standards, medical care and housing for the work force, but it may also be desirable to go much further by initiating programmes for rural community development as a whole (discussed in Chapter 5.7), which also include the establishment of schools, leisure facilities, craft centres and so on.

References

Holopainen, V. (1981) Outline of Finland's Forestry and Forest Policy. Society of Forestry in Finland.

King, K.F.S. (1969) Modernizing Institutions to Promote Forestry Development. The state of Food and Agriculture. FAO, Rome.

Troup, R.S. (1940) Colonial Forest Administration. Oxford University Press.

7 International organizations and conferences

Eero Kalkkinen

7.1 The background

Most forest policy decisions are taken at national level and are determined by national factors, but the international influences on national policies are gaining in importance in a world in which nations are becoming more and more interdependent. We have seen in earlier chapters how this applies to timber imports and exports as well as to environmental problems such as protection against pollution and pests, conservation of wildlife and halting the march of deserts. The question of forestry aid to developing countries is also highly relevant; this too has been mentioned briefly and will be discussed more fully in this chapter.

International co-operation and contacts between foresters were for a long time limited to correspondence among scientists and researchers, usually linked with universities. Such correspondence spread the knowledge of silvicultural and botanical research beyond national frontiers and thus often contributed towards better forestry, but until the first decades of the 20th century, few opportunities existed for foresters to meet physically at international congresses or conferences, and the number of international organizations active in the forestry sector was small and limited to the developed world. After the end of the second world war, however, communications improved rapidly and it became easier for foresters to establish international contacts. Simuntaneously, meeting and conference facilities were created all over the world by the large family of international organizations which came into being at that time.

There are several aspects to the international dimension of forestry: the establishment of a data base, the exchange of technical information, collaborative action including aid programmes and trade are among the more important. The data base relates mainly to the matters discussed in chapters 2 and 3 (forest resource, production, trade etc.) and to the environmental matters discusssed in chapter 4 which generally do not concern forestry alone and are more difficult to quantify (water regimes, climate, wildlife, erosion etc.). The exchange of technical information is well developed. Foresters have few secrets and they are generally very willing to collaborate. Trade negotiations are mostly conducted directly between buyer and seller, but in timber as in other commodities trade also involves broader issues calling for multilateral agreements, which are generally most likely to be reached under the auspices of an international forum or organization. With regard to aid programmes to developing countries, the need in the forestry sector is great and so are the opportunities. Moreover these programmes can also benefit the donor countries and indeed the world as a whole by relieving the pressure both on the world timber market and on the world's remaining tropical forests, the vital environmental role of which was highlighted in chapter 4.

The mechanisms of international co-operation can vary in form, status and mode of operation, they can be governmental or non-governmental, world-wide or regional, global or sectoral, permanent or *ad hoc*, and sometimes vertically or horizontally interrelated. This means that any recommendations, guidelines or decisions designed or intended to be applied by those responsible for national forest policies usually follow the channels determined by the form, status or mode of operation of the mechanism in question. International action may be particular to the forestry sector, or the forestry components may be part of more general activities such as UNESCO's Man and Biosphere Programme, the United Nations Water Conference, Conference on Desertification and United Nations Energy Conference, in all of which the role of forestry has strongly come to the forefront and given foresters an opportunity to present their case to a much wider public.

The following section describes briefly the principles and functioning of the main organizations of relevance to forestry. Three categories are distinguished:

– World-wide governmental organizations
– World-wide non-governmental organizations
– Regional and other groupings

Addresses are given in Annex I.

7.2 World-wide governmental organizations

These are in essence organizations belonging to the United Nations system which are permanently or periodically, directly or indirectly involved in international forestry activities; the programmes and activities of these organizations are controlled by the member governments.

7.2.1 FAO (Food and Agriculture Organization of the United Nations)

This is the most important international organization dealing with forestry. Unfortunately, forestry's share of the FAO budget has been declining in recent years and now accounts for only about 7%. The Forestry Department comprises one Forest Resource Division and one Forest Industries Division, a Policy and Planning Service and an Operations Service. The Department and the member governments are assisted in the orientation of their forest policies by the six Regional Forestry Commissions (Africa, Asia and the Pacific, Europe, Latin America, Middle East and North America). They discuss at their sessions which are normally convened every two to three years, the main issues and problems confronting the forestry administrations of their respective regions; they also formulate recommendations with regard to the orientation of and priorities for FAO's global or regional programmes and activities; not surprisingly, the effectiveness of these Commissions varies considerable. Further advice on FAO's activities comes from the FAO Committee on Forestry (COFO) and the FAO Committee on Forest Development in the Tropics; These consider the findings and recommendations of the regional forestry commissions, and review the implementation of past programmes, as well as those under way; in addition, they also formulate their own, broader conclusions and recommendations for future programmes of work of the organization. While agreements may be reached by the regional forestry commissions or by the two committees mentioned, FAO's own programme of work in the forestry sector becomes effective only after its consideration and approval by the FAO Conference and the various authorities responsible for preparing the organization's programme of work and budget. This institutional framework of FAO makes it possible for member governments to discuss the problems and issues confronting their forest policies and to exchange information on experience obtained in orienting and implementing forest policies elsewhere. It also allows the organization of technical meetings, seminars or symposia, on a regional or world-wide basis, on more narrow sectoral issues that have become important and justify such an exercise. The information contained in the findings of such meetings constitutes one of the most important means of the organization in furthering the exchange and transfer of experience and know-how.

Additional advice and support to FAO's programme of work and activities, affecting also the orientation of national forest policies in many countries, is provided by the three advisory committees, on Pulp and Paper, on Wood-based Panel Products and on Forestry Education, the findings and recommendations of which are important elements for a number of countries in their efforts to improve the industrial utilization of their forest resources, and their forestry institutions. Other advisory bodies for which FAO provides the secretariat include, for example, the International Poplar Commission.

An important tool in assisting the governments in the formulation and orientation of their forest policies is the *World Forestry Congress*. The first two congresses were held before the war (Rome in 1926 and Budapest in 1936). FAO co-operated closely in the organization of the congresses in 1949 (Helsinki), 1954 (Dehra Dun) and 1960 (Seattle), and carries the secretariat responsibility since the sixth congress (Madrid 1966), followed by Buenos Aires in 1972 and Djakarta in 1978. The next congress (originally planned for 1984 but now likely to be postponed) will be held in Mexico. These congresses represent an important international forum at which, at six year intervals, the progress of world forestry and the most important social, economic, technical and environmental problems affecting it are reviewed and discussed by foresters and others directly involved (e.g. planners, economists and industrialists and conservationists).

Since the last World Forestry Congresses, the problems of forest development in the tropics have received growing attention.

The advice and recommendations from all these bodies are taken into account (-some governments believe, not sufficiently-) in FAO's two-year *regular programme of work* and budget which are drawn up by problem area and regional action. The 1982-83 forestry programme contains four major sub-programmes:

1 Forest Resources and Environment;
2 Forest Industries and Trade;
3 Forest Investment and Institutions;
4 Forestry for Rural Development.

The principal objectives indicated under each of these sub-programmes are:

Forest resources and environment
FAO concentrates its advice, assistance and operational activities in member countries on the following sub-objectives:

a The creation, maintenance and updating of a world forest resources data base; assistance to developing countries in the adoption of advanced technologies for forest inventory, monitoring and management, especially designed to suit their purposes and conditions;
b Expansion of forest tree plantations for environmental protection and as renewable sources of energy, timber, fibres, fodder and food, giving special emphasis to drought resistant tree species that can directly support rural development while also providing fuelwood; continuation of programmes of:
  – quality forest seed collection and distribution
  – forest genetic resources conservation and use.
c Development of upland forests in support of erosion contral and water regime regulation by protecting, managing and expanding forest cover in harmony with other land uses in mountain areas; management of wildlife, particularly for its nutritional value, its direct contribution to rural welfare and its significance as part of natural ecosystems.

2 *Forest industries and trade*

a The fundamental objective is to assist member countries to build up the forest industries sector necessary for maximizing the contribution that forest resources can make to economic and social development, especially in rural areas.

b The sub-objectives of this programme are, therefore, to raise the level of self-reliance in the development and management of forest industries and to assist in the orderly development of domestic and export marketing by:

  i the design and application of appropriate technology in the logging, sawmilling, wood-based panels, pulp and paper and wood-based energy industries;

  ii the promotion of investment in the improvement of existing industries in rural areas and the establishment of additional ones, both at the village and larger community levels;

  iii the training of adequate skilled labour forces for logging and forest industries;

  iv the encouragement and expansion of the industrial use of wood-based energy systems in oil conservation and substitution strategies as well as for the improvement of living conditions in rural communities;

  v provision of information on changes in international forest product market supply and demand;

  vi assistance in improved marketing of products from a wider range of wood species and qualities.

3 *Forest investments and institutions*

This programme has as its sub-objectives to:

a keep under review and encourage the development of methodologies appropriate for manpower surveys and training;

b Promote training activities in the forest sector;

c Assist countries, upon request, in institutions building and institutional functions;

d Provide international statistical information with respect to resource, capacity, production, trade and manpower in the forest sector, giving special attention to information gaps related to statistical indicators of rural development in the forest sector.

e Provide a forum for regional discussions of forestry issues;

f Supply information on the development of the sector to both specialized and general audiences;

g Undertake background studies on new issues affecting sector development, which will assist countries in shaping and revising their forest policies and strategies.

4 Forestry for rural development
FAO's advice, assistance and operational activities in member countries concentrates on the following sub-objectives:

a Identification and development of community level systems which enable rural people to best meet those of their needs which are based on forest outputs;
b Strengthening of government and non-government institutions which supprt and service rural forestry activities;
c Establishment of land-use systems which integrate trees and agriculture in a manner which is both environmentally sound and optimally productive for rural people;
d Identification and promotion of off-farm rural employment and income generating activities in forestry;
e Monitoring of fuelwood supply and demand imbalances.

Apart from its regular programme of work described in 1 to 4 above, *FAO acts as executive agency* for the majority of forestry projects financed by the *UNDP (United Nations Development Programme)* which is the largest multilateral agency working in the field of economic and social development. UNDP has financed some 270 forestry and forest industry projects since 1970 at a total cost of nearly 60 million US dollars. While continuing as a financing agency which appoints other agencies to execute projects, UNDP has also established its own small Office of Project Execution to handle special situations; in forestry there have been about one dozen such projects.

FAO also co-operates closely, in forestry-related matters, with the United Nations, its organs and specialized agencies, and with the following in particular: UNEP, UNESCO, UNCTAD, UNIDO, ILO (see particular references to these). In addition, FAO has agreements related to government co-operative programmes with Sweden, Denmark, Norway and with some national governments (Austria, China, Finland, France, India, Italy, USA, USSR, etc.). FAO co-operates in the organization of training courses, seminars, study tours, etc. financed by contributions from these governments and intended primarily for developing countries.

FAO co-operates not only with government organizations and governments, but also with other professional organizations, such as IUCN (Intenational Union for the Conservation of Nature) and IUFRO (International Union of Forestry Research Organizations) on a wide range of subjects.

An important aspect of FAO's work in assisting member governments is a wide range of publications. The topics include forest policies, administration and legislation; education and training; silvicultural practices; logging; watershed management; and other more specific subject matters.

Since its creation, FAO has carried out several studies, either regional or world wide, on trends and prospects in the production, trade and consumption of forest products, the so-called 'Timber Trends Studies', usually in collaboration with the UN Regional Economic Commissions. The continuously improving knowledge of the existing forest resources and their potential, together with generally better and more precise basic statistical information available for each new analysis or study, have made these studies very valuable tools for governments all around the world in their efforts to establish and correctly orient their forestry policies.

7.2.2 UNESCO (United Nations Education, Scientific and Cultural Organization)

UNESCO's foremost activity is its 'Programme on Man and the Biosphere' (MAB), carried out by its Ecological Sciences Division and concentrated primarily on the problems of tropical forestry. Other UNESCO divisions, such as the Environmental Education Division, work in fields related to the subject.

UNESCO action, in co-operation with FAO, is concentrated on supporting and co-ordinating national and regional efforts in environmental monitoring (especially within the international network of biosphere reserves), integrated ecological research in tropical zones, training of scientific personnel and environmental education.

The main focus for MAB activities related to humid tropical forest ecosystems is provided by MAB Project 1, which is concerned with the ecological effects of increasing human activities on tropical and sub-tropical forest ecosystems. The principal objectives of MAB Project 1 is to help develop the scientific basis for the use of natural resources and the management of ecosystems in the tropical and sub-tropical forest zones of the world. Another important aim is to promote self-reliance among countries of the humid and sub-humid tropics in research and management and to encourage the continuing participation of the various sectors of the community in these activities.

Other MAB activities centre, for example, on watershed management in mountainous land.

The basic idea of MAB Project 1 is to develop an international network of field activities that are both integrated and complementary in scope. The intention is to provide a framework whereby the knowledge and experience gained in a particular locality can be transferred and tested in other countries having similar ecological conditions and socio-economic problems. This approach should facilitate the efficient use of scarce manpower and financial resources.

UNESCO's international network of biosphere reserves includes approximately 30 which contain humid or dry tropical forests. These reserves vary in size from 500 to 500,000 ha and they help to conserve biological resources by protecting ecosystems and species.

Technical publications, brochures, reports and audio-visual material serve to inform the general public, politicians, government agencies and scientists.

7.2.3 UNEP (United Nations Environment Programme)

UNEP acts chiefly as a co-ordinating body within the United Nations family on subjects related to the environment.

Although it does not directly carry out projects – a function mainly of the specialized agencies – it does promote world or regional interest in the major environmental subjects and an integrated approach to them. Thus, it co-ordinates and conducts the monitoring of international programmes affecting the environment, which usually require the participation of various specialized United Nations organizations.

UNEP also supports the convening of meetings, workshops, panels of experts and international conferences, as well as the publication of brochures, books and reports on subjects such as desertification, arid land management, alternative energy sources and tropical forests.

Since its creation, UNEP has included in its programme subjects related to forest ecosystems, with special emphasis on tropical forests and arid areas that are becoming deserts.

7.2.4 UNCTAD (United Nations Conference on Trade and Development)

In 1976, UNCTAD adopted a resolution in favour of creating an Integrated Programme for Commodities. Tropical timber is listed among the 18 products in this Programme. Six preparatory meetings on tropical timber have already been held. They examined various methods for improving international markets, including greater stability of foreign exchange income for producing countries and better guarantees of supply for consuming countries. Toward this end, it was suggested that international action should focus on four broad topics: (a) reforestation and forest management: (b) improved knowledge of world markets; (c) increased processing of timber in producing countries; and (d) research and development, including greater use of lesser-known species.

It has also been suggested that UNCTAD and the Common Fund which has been established for commodities might help in implementing the proposed projects, which are co-sponsored by producing and consuming countries.

7.2.5 UNIDO (United Nations Industrial Development Organization)

UNIDO is charged with promoting industrial development in general in developing countries. Its main objectives are a better utilization of resources, the modernization and improved productivity of the existing industries as well as a series of training programmes for improving the technical knowledge and performance of personnel already employed. In the organization of such training activities, through special workshops and seminars in many problem areas, UNIDO benefits from collaboration or co-operation with other international organizations as well as from the development aid of many developed countries. With regard to the forestry sector, UNIDO often co-operates closely with FAO through its participation in basic forest industrial development programmes, (e.g. within the African regional forest industrial advisory group) and direct consultations. Many of UNIDO's efforts are designed to modernize and otherwise improve existing industries, such as sawmills, wood-based panel factories and pulp and paper mills. An important part of UNIDO's programmes in 'the woodworking sector is centred on the establishment and improvement of ssecondary, often small-scale, industries (furniture, prefabricated housing, wooden tools, etc.) which can contribute significantly to economic development at local levels in many of the developing countries.

7.2.6 WMO (World Meteorological Organization)

In 1974, the WMO Commission for Agricultural Meteorology appointed a Working Group on Applications of Meteorology to Forestry. The main outcome was the convening in Ottawa in 1978 of a Symposium on Forest Meteorology. In 1979, the Commission for Agricultural Meteorology established another Working Group on the Role of Forests in the Global Balances of Carbon Dioxide, Water and Energy.

Through the publication of the proceedings of the Symposium on Forest Meteorology, WMO members have been given a fairly clear picture of the requirements for monitoring metereological parameters for application in forestry.

The need for co-operation at international level between metereological and forestry organizations has been accentuated in recent years by the alarming damage to forests attributed to atmospheric pollution (acid rain, heavy metals etc.) in parts of North America and Europe.

7.2.7 ILO (International Labour Organization)

ILO is primarily responsible for:

- ensuring acceptable conditions of employment;
- promoting the safety and health of the workers and improving their social and economic environment;
- many training programmes for improving the vocational performance of the workers and their social and economic status.

In the forestry sector, ILO has been active since the early post-war years in organizing training courses and seminars for forest workers all over the world, often in co-operation with FAO or national programmes for development aid of the developed countries. ILO's contribution is particularly noteworthy within the work of the Joint FAO/ECE/ILO Committee on Forest Working Techiques and Training of Forest Workers, where it shares the secretariat responsibility with the other participating organizations. ILO has also organized special consultations to review the problems and conditions of employment in a number of industrial sectors, including sawmilling, wood-based panel products industries and work in the forests. ILO's activity in these areas is valuable because work in the forest and wood products industries is very often hazardous; the rate of accidents is high and the working environment is difficult particularly in the developing world.

7.2.8 World bank

In 1978, the World Bank published a 'Forest Sector Policy Paper' which outlined a change in its forestry programmes with special reference to the role of forests in rural development and environmental – particularly watershed – protection. The Bank committed itself to increase the total volume of loans in this sector, establishing the target of US $ 500 million for the 1979-83 period, which it has already surpasssed. The Bank is also increasingly concerned about the problem of the energy crisis in developing countries as a resultt of deforestation.

Its forestry and forest industry loans totalling about 1 billion US Dollars have been made to 45 different countries. The bank currently provides about 65% of all external aid funding for forestry and more importantly, it has shifted the emphasis in forestry lending to the extent that rural and environmental forestry investments including fuelwood planting account for more than 60% of its forestry investment portfolio.

In the area of institution building the bank has become more involved in financing forestry training and research and supporting sociological studies of farmer and community level acceptance of rural forestry programmes.

Through its economic, forestry and energy sector review missions, it is assisting member countries to develop forestry and energy strategies, identifying critical policy issues and suggesting appropriate areas for external donor support. In relation to forest policy it is concerned to assist developing country governments to obtain a clear picture of the likely negative human and environment impact of failure to ensure adequate resource commitment to forest conservation and development.

In the supervision of the Bank's Forest Programme, it co-operates with FAO through the World Bank/FAO Co-operative Programme.

7.2.9 *World food programme*

This joint UN/FAO programme has since its creation distributed some 300-400 million dollars' worth of food aid to the developing countries for labour intensive projects intended to promote economic and social development and employment. Many of these projects have forestry components and thus assist in the implementation of the receiving governments' forestry programmes, e.g. plantations for soil, water and general environmental protection, creation of small forests near villages, windbreaks and other forestry work for rural development. In arid and semi-arid zones, forestry components of the projects emphasize environmental functions, such as dune stabilization, the fight against erosion and desertification, shelter strips and windbreaks.

7.3 World-wide non-governmental organizations

7.3.1 *IUFRO (International Union of Forestry Research Organizations)*

IUFRO was established in the years 1890-92 and is thus the oldest international forestry organization. Until World War II its activities were confined almost entirely to Europe. Membership then spread to developed countries on other continents and more recently many forestry research institutes in developing countries have also joined bringing total membership to over 500 institutes with about 10,000 scientists. IUFRO's principal objective is to promote the exchange of information on the results of research and the co-operation on specific problems. Its structure consists of an Executive Board and six technical divisions, namely:

1 Forest Environment and Silviculture;
2 Forest Plants and Forest Protection;
3 Forest Operations and Techniques;
4 Planning, Economics, Growth and Yield, Management and Policy;
5 Forest Products;
6 General Subjects.

In the absence of central funds, the basic work is financed and carried out by the members of the different divisions or their sub-groups, in the form of agreed research projects, technical meetings and exchanges of information; in addition, IUFRO convenes every five years, world congresses for its member institutes (the last, the XVII Congress, was held in 1981 in Japan). These congresses provide a forum for the review of forest research work in the world and give guidance for the orientation of further work.

A number of international organizations co-operate with IUFRO especially

FAO which is represented at IUFRO's Executive Board, while representatives of IUFRO participate in the work of all FAO's forestry bodies. IUFRO's increasing concern with the forestry problems of developing countries is dealt with in Chapter 6.5 on research.

7.3.2 IUCN (International Union for the Conservation of Nature and Natural Resources)

This organization has over 400 members all over the world, including governments, official agencies and national and international government related organizations, such as UN, FAO, UNEP and UNESCO. The major objective of IUCN is to conserve and promote the diversity of the biosphere within a rational management of the earth's natural resources. The World Conservation Strategy, sponsored by UN, FAO, UNEP, IUCN and WWF (World Wildlife Fund), maintains that conservation will only be possible in the long term if and when basic human needs are met. IUCN has elabaorated national conservation plans for several countries, particularly in tropical South-East Asia and in Latin America. A number of its field projects are specifically related to tropical forests, mostly with the objective of establishing or maintaining protected areas. In this respect, it also assists governments all over the world in matters concerning National Parks and other protected areas.

7.2.3 WWF (World Wildlife Fund)

WWF is the largest private international conservation organization and is devoted mainly to the protection of endangered species and their habitats. Its principal activities are:

– basic and applied research for the promotion of new reserves and biological parks;
– aid for the maintenance of those already in existence.

WWF also serves as an intermediary body between governments and private interests.

7.3.4 ICRAF (International council for research in agroforestry)

This Council which was created in 1979 promotes, as its name implies, the combination of agricultural and livestock production with forest production. The object is to improve living standards in developing countries. ICRAF coordinates and participates in the development and application of agroforestry in both deforested and forest zones.

7.4 Regional and other groups

7.4.1 United Nations regional economic commissions (ECE, ECA, ECLA, ESCAP, ECWA)

The Economic Commission for Europe (ECE) has, throughout its existence, devoted a considerable part of its programme to the problems of production, trade and consumption of forest products. Its periodic medium and long-term studies and projections of trends in the forest products sector, in co-operation with FAO, have helped to guide the forestry administrations of the region in the orientation of their forest policies. The operational mechanism of the ECE consists of its *Timber Committee*. This reviews annually developments and trends in the region's forest economy; and under it, special meetings, seminars and symposia are organized on subjects which justify special attention. The Timber Committee also has two subsidiary bodies, the ECE/FAO Working Party on Forest Economics and Statistics and the FAO/ECE/ILO Committee on Forest Working Techniques and Training of Forest Workers. The activities of all these bodies have influenced or contributed to the orientation of forest policies also in many countries outside the ECE region proper, which, in addition to Europe, includes North America and the USSR. The ECE Timber Committee and its subsidiary bodies constitute a good example of co-operation between international organizations.

The Economic Commission for Africa (ECA), although without the organizational structure of the ECE, nevertheless devotes (in collaboration with FAO) part of its activities to the problems of forestry and related areas in its region. ECA acts, for instance, as executive agency for the regional project on Development and Conservation of Forest Resources, which is primarily concerned with tropical forestry. The main subjects already studied or currently under study are conservation of resources, role of the forestry sector in national economies and the exploitation and management of non-wood forest resources. The project is intended to stimulate co-operation to solve conservation problems. Its initial monitoring work and evaluation of vegetal cover will enable ECA to warn the governments and authorities concerned of the changes occurring and to advise them on measures that should be taken to arrest ecological damage, so that they may orient their forest policies accordingly.

The Economic Commission for Latin America (ECLA) has devoted itself mainly to industrial development and the study of trends in the production and consumption of wood and its derivatives; it has done so in collaboration with FAO through the FAO/ECLA/UNDP Advisory Group on Forest Industries. A joint study on Latin American Timber Trends and Prospects was intended to guide the region's forestry authorities in the formulation and orientation of their forest policies.

The Economic and Social Commission for Asia and the Pacific (ESCAP) and that for Western Asia (ECWA) have shown a considerable concern for the problems of deforestation and desertification in their regions and have drawn the attention of their member governments to action taken and recommendations formulated by other international bodies and conferences, e.g. the United Nations Conference on Desertification.

7.4.2 OECD (Organization for economic co-operation and development)

OECD, formerly OEEC (Organization for European Economic Co-operation) today groups the majority of the world's industrialized non-socialist countries. Although its programme of work and activities do not permanently or continuously deal with forestry, there are several aspects of its work that are connected, directly or indirectly, with forestry related topics and, hence, with the forest policies of its member countries. The OECD Industry Committee has an expert group on pulp and paper, which meets periodically and reviews the developments of this sector of the economy in its member countries. OECD also published statistics on pulp and paper and on this sector's raw material needs. Recently its work has tended to be on specific problems within the pulp and paper industries, such as energy and pollution, rather than on these industries as a whole. In 1974 the OECD Committee for Scientific and Technological Policy set up an *ad hoc* Group on Materials Resources, which decided to undertake a case study on wood as a renewable material, after having completed a study on copper, a non-renewable material. This report, published under the title 'The Life Cycle of Wood', aims at the identification of the 'principal points of leverage within the system where successsful materials research and development can be expected to contribute the most to meeting anticipated supplies needs for modifying demand . . . or to addressing particular national objectives sought over a given time.' In its conclusion the study formulates proposals and recommendations for research and development in the forestry and related sectors.

7.4.3 EEC (European Economic Community)

This European organization which is based on the Treaty of Rome (signed in 1957) has as its main organs the Council of Ministers, the European Parliament, the Court of Justice and the Commission which is the central administrative and executive organ. At present the EEC groups ten countries (Belgium, Denmark, Federal Republic of Germany, France, Greece, Ireland, Italy, Luxembourg, the Netherlands and the United Kingdom) with two others (Portugal and Spain) waiting for admission. Forestry and related subjects were not included as major issues in the original Treaty of Rome, and have therefore carried relatively small weight in the activities of the EEC. A Forestry and Environment Division is included in the Commission's Directorate General for Agriculture. The main forestry activities are as follows:

- the co-ordination of national forest policies of the Member States;
- the financing of specific forestry measures especially in the Mediterranean region of the Community where existing programmes provide for a community contribution equivalent to some 50 million US dollars per year;
- legislative measures to prevent the spread of tree diseases and to ensure quality standards of forest reproductive material;
- the preparation of studies on specific problems and issues such as public access to forests, forestry taxation and incentives and mechanization of forest operations;
- a programme of forest and forest products research; this programme contributes the equivalent to some 3 million US dollars per year to the financing of research projects of Community interest and for the better co-ordination of research undertaken nationally; states which are not members of the EEC may also participate in this programme;
- forestry projects are included in the Community's aid programme to developing countries. A few of the larger ones involve an EEC contribution of several million dollars.

The Community's forestry activities are kept under review by the Chiefs of the forest administrations of the member states and by various groups of experts who meet periodically at the invitation of the Commission.

7.4.4 Regional development banks

These banks, which have been established by the governments of the regions with the participation of financial or banking institutions and governments of the developed countries, have gradually been giving increasing attention to the forestry sector through granting loans for forestry development and advising the governments in this regard.

- *The African Development Bank* has so far done little for the forestry sector. The need for appropriatae investment programmes and projects is great, but neither the Bank nor the governments have given these the priority they deserve;
- *The Interamerican Development Bank* has shown an interest in the forestry sector in Latin America and granted considerable development loans, destined mainly to forest industries. In its document 'Operational Policy: Forestry Development', the Bank reflects its willingness to finance forestry projects either public or private, in the following fields: development of natural re-

sources; establishment of new plantations for industrial use; establishment of industries for the utilization of natural forests and new forest plantations; construction of access roads to forest zones and acquisition of equipment for logging and transport of products; and regeneration of natural forests or establishment of new forest plantations for multiple use (watershed protection, recuperation of eroded soil, national parks etc.).

- *The Asian Development Bank* has so far granted only a small percentage of its loans to the forestry sector, but this is likely to change in view of the serious problems affecting the forest resources in the region and the great opportunities offered by their rational development.
- *The Caribbean Development Bank* which was established after the regional banks mentioned above, has also begun to show an interest in forestry projects.

7.4.5 Other regional groups

Apart from organizations which cover principal geographic regions or a combination of these, there are groups active in the forestry sector which cover a given sub-region or groups of countries having historical links; similar environmental, economic and social conditions; and in which the principal objectives of forest policies have many common aspects. Two such groups, although not constituting an institutional and rigid framework for co-operation, can be mentioned as an example of a useful, limited system of contacts between countries and having the possibility of influencing forest policies of the countries participating in the system. One is the *Nordic Forest Union*, which groups the five Nordic countries (Denmark, Finland, Iceland, Norway and Sweden). Its objective is to organize co-operation on forestry matters of common interest to the member states. Periodic congresses of the Union review common problems and strengthen the ties among the participating foresters.

Another association is that of the central-European alpine countries (Austria, Federal Republic of Germany and Switzerland) the wood research institutes of which organize periodically the so-called *'Three Countries Wood Meetings'* to review and discuss together the special problems which are common to their conditions and situations.

The Confederation Europeanne d'Agriculture (CEA) is an example of a regional non-governmental organization for agriculture and forestry. Its members are agricultural and forestry associations from the whole of Western Europe. Commission V of the CEA deals specifically with the forests in private ownership and in the ownership of local authorities (communes etc). The CEA's activities include annual meetings of the members and the examination of specific matters of common interest by study groups; it maintains a permanent secretariat at Brugg in Switzerland. Other organizations in Europe which have shown an interest in forestry are the *Council of Europe* in Strasbourg and the *Centre for European Policy Studies* in Louvain-la-Neuve in Belgium, both of which have commissioned studies on forest policy in Europe.

Another type of international co-operation is that which takes place among member countries of the *British Commonwealth*. Its activities are centred around:

- the Commonwealth Forestry Insitute which was founded at Oxford in 1924; it is engaged in overseas research and education and it also provides other services which are now available to countries outside the Commonwealth as well as to those within it:
- the Commonwealth Forestry Bureau; this is part of the world-wide information service of the Commonwealth Agricultural Bureaux, which publish Forestry Abstracts and Forest Products Abstracts;
- the Commonwealth Forestry Conference, held every five or six years to provide a forum for the discussion of forestry problems and achievements in the countries represented, many of which, despite their diversity, share English as a common language and a similar approach to forest policy and administration.

While all the examples given above involve the participation of forestry institutions and administrations from several countries, one should not forget the important bilateral programmes of development aid which today are part of the relations between the developed industrial countries and those less favoured and still at an early stage of development. All these programmes include, to a variable degree, assistance to forestry and related sectors according to the needs indicated by the receiving countries. This assistance, which is sometimes provided in co-operation with international organizations such as FAO, ILO, UNIDO, may cover a great variety of activities among which training and education figure prominently, either through the granting of fellowships in developed countries, or the establishment of educational and training centres in the receiving countries, or the organization of seminars and courses on specific subjects, for participants from a single or several developing countries. The size of the forestry component in the development aid programmes often depends on the importance of the forestry sector in both the donor country and the country for which the aid is entended. The political and trade advantages which a donor hopes to gain as a result of the aid may also influence the issue.

7.5 Problems and achievements

International activities in forestry have developed rapidly from very small beginnings during the past thirty years and it will be at least another generation before the full effects of what has been done up to now will become apparent. The seeds sown by international co-operation may take a long time to germiniate. In order to co-operate it is first necessary to communicate and in order to communicate one must understand one another's circumstances, aspirations and points of view. This all takes time, but there can be no doubt that at the human level a good foundation for the future has been laid. The professional horizon of foresters almost throughout the world is no longer restricted by national frontiers.

There have also been some very notable practical achievements. Forestry statistics and forecasts of supply and demand of timber products are now available on a world basis when a generation ago there was almost nothing; most countries in the world now have forest services which provide at least a starting point for the rational development of a country's forest resource; and perhaps most important of all there is a free flow of forestry know-how and expertise between countries. There is, of course, room for further improvement in international co-operation in forestry and it may help to bring this about if, with the benefits of hindsight, we examine what might be done better in future.

We have seen that, in addition to FAO which is the principal agent for international co-operation in forestry, many other international agencies and organizations have entered the forestry sector either through the creation of new types of development assistance or because certain forestry aspects were part of their general activities. Inevitably, there were some shortcomings and even some failures for structural, political, economic and technical reasons. Efficiency is obviously more difficult to achieve in an international organization than in a national one because the staff do not have a common mother tongue and they are accustomed to different ways of working. There has also been some 'empire building' which has led to unnecessary duplication of effort, but it must be remembered that not all duplication is wasteful. For example, much has been gained by FAO and IUFRO dealing with the same topic in collaboration with each other. The national participants at meetings convened by FAO generally speak for their governments and can get things done; the participants at IUFRO meetings (sometimes the same experts) speak only for themselves and can therefore express themselves more freely. Organizations tend to become more bureaucratic as they grow bigger; international organizations are no exception and there may indeed be a danger in some that too large a proportion of personnel and funds are devoted to the central administration and too small a proportion to the operations for which the organization was established; and that administrative convenience rather than operational efficiency influences procedures and decisions. In short: 'the administrative tail starts to wag the operational dog'. The main remedy for most of these problems lies in firm control by the member countries; this control is admittedly difficult to achieve, because it is relatively easy for an international agency to play its Member governments off against each other.

Development aid creates some special problems for donors as well as for the recipients. Programmes often bear little relation to a country's priorities or to each other. Common sources of this inefficiency are rivalries which occur between the various international and bilateral aid agencies as well as between their counterparts in different ministeries in the receiving countries of the development aid. As a result some forestry programmes or projects have escaped from the control or even knowledge of the central forestry administration. The institutional weakness of most forest administrations in the receiving countries (usually part of the ministry of agriculture) has made it difficult for them to compete with other ministeries such as those for planning and finance, industry and trade, or with special development authorities, which usually are stronger, both institutionally and politically.

One way of trying to achieve a better co-ordination of effort is for the national authorities and the international and bilateral donor agencies in a country to meet periodically in order to inform each other of their actions and intentions;but that will only work if all participants are prepared to lay their cards on the table; unfortunately vested interests on both sides may prevent this from happening, e.g. a national authority may not wish to divulge that it has approached more than one potential donor for a particular project or a donor agency may not wish to reveal that it has entered into discussions about a particular aid project for fear that another agency might intervene; that risk applies particularly in the case of prestige projects.

Bilateral aid has sometimes been less effective than had been hoped for, either because the equipment which the donor could offer was unsuitable for the conditions of the recipient country, or because there were ulterior political or trade motives. The interests of donors and recipients of aid are not necessarily incompatible but recipient countries would do well to look into these possibilities when they negotiate aid agreements.

A few developing countries have run into difficulties by accepting more aid programmes than they can 'digest' with their limited resources of highly qualified professional staff: by having to divert too many forest officers to service new aid projects, work already in progress has suffered.

Another problem has been the selection of experts for the development programmes. Originally, most of these came from developed countries, were highly qualified professionally, and did a good job. Some, however, failed to adapt their know-how, methods of working and of behaviour to the unfamiliar physical and social environment of the receiving country. In some instances, highly qualified experts were recruited for jobs which required competent general practitoners; this was frustrating for the experts and inefficient: the experts could not make use of their expertise because of lack of local facilities and some of the work that needed doing could not be done because it was outside the experts' special field of knowledge. Some experts have also caused frustration among their national counter-parts by under-estimating their professional standards and the value of their knowledge of local conditions. The main lessons to be learnt from these experiences seem to be twofold: first, that temperament, character and commonsense are as important as professional qualifications, and secondly, that an expert only becomes effective after having worked for at least a year or two in a country and got to know its language, its people and its forests. On the other hand an expert who is kept in one country for too long may get 'stale'. Ideally, a period of service on development aid will alternate with a period of 'recycling' in the expert's home country.

In recent years, aid agencies have made fuller use of the expert potential available in the developing world. Problems similar to the ones described above may arise also here except that it may be easier to obtain experts who are familiar with life and work in a poor country. It has been suggested that the relatively few highly qualified foresters in a developing country, should not be tempted to work abroad because they are needed so badly at home, but it would be unfair to deprive such people of opportunities to gain additional experience by working in other countries. Moreover, this extra experience may benefit their own country to which many return eventually, enriched by new ideas and a broader outlook. To oblige them to return would, however, rarely be either practicable or even desirable.

The employment of expatriates is always expensive because in addition to the usual salary there are all the costs and allowances associated with work in a foreign country: travel, housing, schooling of children etc. In the early 1980's the cost of keeping an expert in post averaged nearly 100,000 dollars per year. It pays to reduce the number of foreign experts to a minimum, to be highly selective in their recruitment and to ensure that those who are recruited are given every opportunity to develop their full potential.

In conclusion, the best way forward would seem to lie in seeing both the past achievements and shortcomings in proper perspective. Extreme voices have been heard suggesting that the whole existing system of international co-operation and aid should be dismantled and a fresh start made; at the other extreme, the shortcomings are glossed over with a pretence that all is best in the best of all possible worlds. Both these approaches appear equally inadvisable: the first, because it might lead to the destruction of what has been achieved, the second, because it would at best put a stop to progress, and at worst lead to complete disaster.

With all its shortcomings international co-operation in forestry has already resulted in worthwhile achievements upon which we can build. Let us do so keeping the shortcomings of what has happened clearly in view so that we may do better in the future.

Annex

Addresses of international organizations

7.1.0   *United Nations Organization*
United Nations Development Programme (UNDP)
New York 10017, USA

7.1.1   *The Food and Agriculture Organization of the United Nations (FAO)*
Via delle Terme di Caracalla
00100 Rome
Italy

7.1.2   *United Nations Educational, Scientific and Cultural Organization*
(UNESCO)
Place de Fontenoy
75007 Paris 7e
France

7.1.3   *United Nations Environment Programme (UNEP)*
PO Box 30552
Nairobi
Kenya

7.1.4   *United Nations Conference on Trade & Development (UNCTAD)*
Palais des Nations
1211 Geneve
Switzerland

7.1.5   *United Nations Organization for Industrial Development (UNIDO)*
Vienna International Centre
PO Box 300
Vienna 1400
Austria

7.1.6   *World Meteorological Organization (WMO)*
41 avenue Giuseppe Motta
1211 Geneve 20
Switzerland

7.1.7   *International Labour Organization (ILO)*
4 route des Morillons
1211 Geneve 22
Switzerland

7.1.8   *International Bank for Reconstruction & Development (IBRD)*
World bank
1818 H St NW
Washington DC 20433
USA

7.1.9   *World Food Programme (WFP)*
Via delle Terme di Caracalla
00100 Rome
Italy

7.2.1   *International Union of Forestry Research Organizations (IUFRO)*
Schönbrunn
Tirolergarten
1131 Vienna
Austria

7.2.2   *International Union for the Conservation of Nature and Natural*
Resources (IUCN)
1110 Morges
Switzerland

7.2.3   *World Wildlife Fund (WWF)*
Avenue du Mont Blanc
1196 Gland
Switzerland

7.2.4   *International Council for Research in Agroforestry (ICRAF)*
Nairobi
Kenya

7.3.1   *Economic Commission for Europe (ECE)*
Palais des Nations
1211 Geneve
Switzerland

Economic Commission for Africa (ECA)
PO Box 3001
Addis Ababa
Ethiopia

Economic Commission for Latin America (ECLA)
Edificio Naciones Unidas
Avenida Dag Hammarskjold
Vitacura
Santiago, Chile

Economic and Social Commission for Asia and the Pacific (ESCAP)
United Nations Building
Rajdamneon Avenue
Bangkok, Thailand

Economic and Social Commission for Western Asia (ECWA)
Nabil Adel Building
Bir Hassan
Beirut, Lebanon

7.3.2   *Organization for Economic Co-operation & Development (OECD)*
2 rue Andre Pascal
75775 Paris Cedex 16, France

7.3.3   *European Economic Community (EEC)*
200 rue de la Loi
1049 Bruxelles, Belgium

7.3.4   *African Development Bank*
BP 1387
Abidjan, Ivory Coast

Interamerican Development Bank
808 17th Street NW
Washington DC. 20577, USA
Asian Development Bank
PO Box 789
Manila, Philippines

7.3.5   *Commonwealth Forestry Conference*
The Secretary, Standing Committee on Commonwealth Forestry
231 Corstorphine Road
Edinburgh EH12 7AT
Scotland

Commonwealth Forestry Institute
South Parks Road
Oxford
OX1 3RB, England

8 Policy formation

Fred Hummel and Adriaan van Maaren

In the foregoing chapters of this book the main issues of forest policy and the relevant options are discussed by one: the role of forests and related kinds of land use in national life, the way in which forest products and services can best benefit people both in the short and long term, the possible responses to the increasing pressures which threaten the forest estate in so many countries and, finally, the institutional and administrative framework at national and international level without which sensible policies cannot be developed and implemented.

While drawing on the elements previously discussed in this book, this chapter is concerned with the *process* by which forest policy is formed, implemented, reviewed and developed in the light of changing circumstances.

The *first* (8.1) part outlines in general terms the process of policy formation and the various factors to be taken into account in this process. Clearly, this broad, somewhat *theoretical framework* must in practice be adapted to the special constellation of circumstances of each country. A country with a large forest resource geared to timber production presents different problems from a densely populated industrialized country with few forests which are primarily needed as 'green belts' and for recreation.

In the *second* part (8.2) the process of policy formation is described in a more *pragmatic* way; specific options are discussed and examples given from practice. The emphasis here is on the dynamics of change, on the need for a forest policy not be static but to be constantly reviewed and updated so as to meet the changing needs of society in a changing world.

8.1 The framework

8.1.1 General points

Policy formation is above all a government responsibility because the implementation of important decisions generally requires action by the government or at least the approval of government. It is only within the framework of general government policy for forests and forestry that meaningful decisions can be taken by woodland owners, foresters and others concerned. Usually the hierarchy of policy decisions is from the centre outward and from the top downward but the flow of information and views which precede these decisions should be also in the reverse direction. In countries where society has not yet developed sufficiently to enable an effective two way flow of information and views to take place, it is decision makers who must ascertain as best they can, preferably by visits to the field and discussions with the people on the spot, what the facts, views and local interests are.

The objective must be to reach *realistic* decisions which can be implemented and enforced. A mere statement of what is desirable or of what is likely to happen is no policy and is at best of limited use. To be realistic, a policy decision must be:

- appropriate to the physical, social and economic conditions of a country. Climate, topography, extent and type of forest cover, density of population, degree of economic development and systems of land tenure are among the factors to be considered; furthermore, the analysis of these basic factors should be future-oriented, taking into account foreseeable changes;
- acceptable to or at least tolerated by those who are affected by it; experience in many countries has clearly demonstrated that no laws and administrative measures can secure the implementation of policies which threaten the customs or livelihood of the people affected especially those living in or near forests (for example see section 8.2); policies are more likely to be accepted if they are explained and if there has been prior consultation;
- backed by the necessary technical, financial and human resources for implementation as well as by legislation; it is useless to will the ends without willing the means.

Policies are designed to benefit people; but whom? There may be conflicts of interest between the population as a whole and those living in or near forests or depending on forests and forest products for their living, especially woodland owners and all who work in the forest or wood processing industries. It is in the poorest countries where the majority of people live in rural areas and depend entirely on forest produce for many necessities of life, including wood fuel to cook their food. The interests of these people should be paramount although they usually have little political influence.

It may also not always be easy to reconcile the achievement of immediate benefits with maintaining and if possible improving the long term production potential of forests. The individual is generally mainly concerned with the benefits that may accrue during his life time, the politician with what happens before the next election. Foresters on the other hand are all too prepared to ignore the immediate futue in favour in the next century.

Forest policies and other policies must be consistent with each other and where there are conflicts of interest, these must be resolved. As stated in chapter 1, forest policy is an important component of land use policy, in the absence of which it is difficult to develop a sensible forest policy. There are also links with other policies: those concerned with agriculture, the environment, industry, trade, employment, regional development, to name some of the more important. What is however, crucial is that forest policy must be developed as a coherent whole instead of as a series of appendages to these other policies. It is only when forest policy is developed as a sector policy in its own right that the various aspects of forest policy can be properly co-ordinated. The co-ordination of policies is greatly facilitated within the framework of *regional* policies because every region has some specific characteristics and problems.

Finally, forest policy formation is a continuing process. Policy must be adjusted in the light of changing physical and economic circumstances and changing demands by the various sectors of society. It is, however, a great advantage if forest policy is not made the subject of party politics. Forest policy must be progressive, but frequent changes of direction are bound to harm any long term enterprise such as forestry.

8.1.2 Responsibilities

Although the more important decisions are generally the responsibility and prerogative of the central government of a country, some national policy decisions are also taken by others and the circle of those who participate in the progress formation is even wider.

The respective roles of central government, regional or local governments, institutions and groups representing particular interests depends largely on the distribution of power in a nation; thus in countries with a federal structure the role of the individual states or provinces will be greater than in countries with a more centralized government structure. The influence of particular interest groups such as woodland owners and forest workers trade unions will depend on the efficiency of these organizations as well as on the role assigned to them by law and custom.

The main responsibility for actually doing the preparatory work, i.e. the assembly and analyses of relevant information, the identification of problems, options and goals, and the organization of consultations with all concerned is generally assigned by government to the forest authority, or to some committee or group of experts specially designated for the purpose. Such committees or groups may consist entirely of officials from various government departments or of external experts appointed as consultants or of a mixture of both. Specific examples are given in section 8.2. The most appropriate solution will vary according to circumstances.

Whatever the administrative set up, the forest authority has a crucial responsibility for advising government on policy.formation on the one hand and for acting as an executive arm of government for the implementation of policy on the other hand.

An important dividing line is between *government* policy on the one hand, and *management* and *administration* on the other. As already stated, the provision of the financial, technical and human resources required for the implementation of policy decisions should not be separated from the policy decisions themselves which otherwise can at best be described as pious hopes rather than as policies. But the effective deployment of these resources requires plans, programmes and budgets first at national and then at more local levels. At national level these measures are clearly part of, or at any rate closely linked with, policy formation while at local level they are mainly a matter of management and administration. The distinction between policy and management depends less on the content of a decision than on the level at which the decision is taken. Thus the methods of sale of timber are a matter of policy if they are prescribed by government, but they are a matter of management if left to the discretion of the forest owners, public and private. Forestry in this respect inevitably follows the trend of other sectors of the economy. Where government intervention is strong, forestry is unlikely to escape; conversely, where intervention is kept to a minimum, forestry will be left to manage its own affairs except where there are good reasons to the contrary. As forestry in most countries can claim only a very limited amount of senior politicians' time, government decisions cannot easily be changed if they are no longer relevant. Attention to detail also tends to detract attention from more important issues. There is thus a strong case for limiting government policy decisions to major issues and leaving the implementation to the forest authority, although this may not always be welcomed by the forest authorities themselves, some of which prefer to operate by a 'rule book' rather than to take decisions and be held responsible for them. This of course raises the much wider policy issue of the role and training of government officials.

8.1.3 Public involvement

Policies are mainly about people and for people; it is a common pitfall to think only of the forest resource when developing forest policy. It is people who have to make the policy work and who are intended to benefit from it. Social factors and local customs must therefore not be ignored. In fact social considerations may have to outweigh technological considerations. Various components of society have different interests and make different demands which may be in conflict with each other and may exceed the limits set by the forest resource. In particular, it is essential to differentiate between on the one hand the demands of the people who live in or near forests and depend directly on the forests for their necessities of life and their work and earnings and on the other hand the rest of the population whose links with forests are more indirect; they consume forest products, enjoy walking in forests and they benefit from the other environmental functions of forests but may not even be aware of these benefits.

Policy formation must therefore start by identifying the various demands on the forest by various sectors of society and then seek to reconcile these demands with each other and with the forest resource and the human and financial resources available to develop it. In part, the reconciliation is helped by the application of *forest science* and other related applied sciences because this will maximize the products and services which forests can provide on a sustainable basis. Bridging the almost inevitable remaining gap between the various demands and what the forest can supply is a matter of political decision on priorities. Unfortunately too little priority tends to be accorded to the interests of people in rural areas who depend most on forests but generally have less political influence than the people in towns.

Insofar as dialogue of the forestry and forest industries sector with the public is concerned, three categories of the latter may be recognized:

– The general public. Everyone benefits in one way or another from the forest. This dialogue between the forest and forest industries sector and the public should have, as a primary objective, that of making the public generally more aware of all the marketable and non-marketable benefits it receives from the forest resource; but listening to public opinion is equally important.
– The opinion formers. The public includes those groups and individuals who feel, with or without justification, that they have the competence or authority to express opinions on the role of the forestry and forest industries sector and on the way it should be managed and used. The 'opinion formers' include the media, schools and colleges, authors and artists, youth clubs and organizations, leaders of workers' unions and so on.
– Policy-formulators and decision-makers. The third group of the public consists of the legislators of all levels and aspects of government (planning, financing, taxation, etc.) The forestry and forest industries sector is only one among many with which they must deal. They are not necessarily better informed than the general public about that particular sector. It is therefore very important to provide them at the right time with the proper information.

Foresters may have a tendency to take an over-simplified view of the public, both with respect to the diversity of the groups with which they have to deal and to their sophistication, particularly of the opinion-formers. A more conscious effort to establish a dialogue with them would undoubtedly be in the interests of all concerned. The forestry and forest industries sector must take the initiative in putting the facts firmly but objectively before the public; waiting to react to criticism or adverse opinion is ineffective.

The decision on the optimum combination of the particular functions of forests is a political concern and therefore one of the main topics of forest policies, but political decisions must have a sound motivation induced by experts, in this case by foresters. If foresters want to improve the existing forest policy in a given country, they have to put forward a comprehensive conception of the optimum use of forest resources. If foresters have such a conception, they can influence both decision-makers, ministerial authorities, and opinion formers and even the general public in the suggested direction.

8.1.4 The steps in policy formation

The crux of policy formation is the definition of objectives or goals. The choice of goals is limited by physical constraints (land, money, etc.) and by certain *a priori* guidelines or principles or priorities which have either been specifically laid down by the government or are implicit in its general policies. While the goals to some extent determine the information needed to identify options, it is also true that once the facts have been assembled new objectives and new options may suggest themselves. We shall now consider in turn the various steps in policy formation; they are similar to those in any planning operation and indeed the dividing line between policy formation and planning is somewhat arbitrary.

Assembly of information. The starting point for policy formation is existing forest policy. One cannot reach sensible decisions on where to go without knowing from where one is starting. Even in countries with a long forest history, forest policy is not always succinctly defined but has to be deduced from a host of legislative and administrative measures taken individually and over a period of years or even centuries. In countries without a forestry tradition, there may be no forest policy as such, but there are bound to be laws and customs affecting the use of forest land and the exploitation of forest products which must not be ingnored.

The second category of information to be collected refers to the forestry situation: forest areas, ownership, standing volume, increment and volume of cut as well as all the relevant facts on climate, topography, soils, fauna, flora etc. Ideally a comprehensive national forest inventory should be undertaken, but there is a risk that by trying to get all the desirable information, essential policy decisions may be delayed. The quest for the best may become the enemy of the good.

Analysis of options and the process of consultation. Before options can be analysed and the advantages and disadvantages of each considered, the options must be clearly identified. This may sound abvious but is often not done. This omission may be due to insufficient experience of planning methods by those concerned or the omission may be intentional: for reasons that may be good or bad, but usually bad, governments sometimes do not wish to spell out all the options.

The analysis of options should take into account not only the relevant facts concerning the forests, laws, and customs, but also the views of all interested parties because Government decisions on forest policy are almost bound to affect many interests not all directly concerned with forestry. In the first place, there are the various ministries and departments of government; directly responsible for forestry and other rural land use as well as those concerned with finance, environment, employment, trade and industry, and possibly others. Then there are all the non-governmental organizations, such as those representing private woodland owners, emloyees, forest industries and so on. In some instances consultation within government circles is complicated by the fact that different ministries may be responsible for different aspects of forestry. That such a state of affairs is generally undesirable has already been explained in section 6.6. It is, however, equally undesirable if, as has sometimes happened, ministries with an interest are not consulted. There may also be organizations outside forestry which may understandably wish to be consulted for example organizations representing various environmental, hunting or tourist interests. The weight to be attached to such lobbies is a matter of political judgement, but they should at least be heard.

The effectiveness of consultation with non-governmental interests depends essentially on two factors, First, there must be in existence organizations representing these various interests; otherwise there cannot be a meaningful, two way flow of information and views. Secondly, there must be provision for bringing the leading representatives of these organizations together for joint discussions with government, because it is as important that these various interests should learn to understand one another as it is for them individually to be in touch with government. Many developed countries have this type of consultative machinery but the details vary; in the Federal Republic of Germany for example, there are two major consultative bodies, one for forestry (Forstwirtschaftsrat) the other for wood and wood processing (Holzwirtschaftsrat) while in Britain growers, wood processing industries and the timber trade are all brought together in the Home-Grown Timber Advisory Committee. In developing countries, organizations of these different sectors often do not exist, and may be difficult to establish, but it is always possible to ascertain the views of a representative sample of individuals of these different sectors, and this is essential if policies are to be realistic and acceptable. To be effective these contacts should be made where these people work and live and not in a government office.

What may happen in both developing and developed countries, is that some sectors are much better organized than others, and are thus in a better position to exert influence. A common example, is that the sawmilling industry consists of a very large number of small units and is poorly organized, while the pulp and paper industry consists of a few larger units, which are obviously easier to bring together. Under these circumstances, governments must take care not to listen only to the loudest voice.

Definition of objectives. As already mentioned, a broad definition of objectives is essential before even the facts of the situation, in so far as they are not already known, are assembled, but the analysis of the options revealed by the facts may suggest some modification of the objectives. For example, a prior objective may be to grow timber for export as a means of earning foreign exchange. The analysis of the situation may reveal that the pursuit of this goal might make it impossible to meet domestic demand for wood and that it may, therefore, be preferable to give priority to the meeting of this domestic demand. For examples of objectives see 8.2.

A common mistake is to list a series of objectives without any consideration of priorities, of mutual compatibility or of the resources that are needed for their implementation. It is true that policy objectives need not be – and indeed usually cannot be – defined as precisely as the management objectives of a small commercial enterprise; nevertheless vagueness or a multiplicity of objectives leads to inaction or chaos or both. It is of course possible to state that different objectives should be given priority under different circumstances or in different parts of a country. Thus timber production for industry may be top priority in some regions, forestry to meet the domestic needs of villagers elsewhere, protection forestry in certain water catchment areas in the mountains, forestry for recreation near towns or where tourists are important. It may also make good sense to differentiate objectives by ownership; for example, that private woodland owners should, within certain contraints, be allowed to pursue their own interests and production goals while the service functions are given higher priority in publicly owned forests. However, if objectives are spelt out in too much detail, they will either be ignored or rob those who have to implement policy of the power of initiative, which is so essential if policies are to be adapted to the circumstances of a particular forest or forest owner.

The taking of decisions. Decisions on forest policy are often taken individually rather than as a comprehensive 'package'. There are good reasons for such a 'piece meal' approach. It may be that a particular issue requires immediate decisions, sometimes for reasons completely outside forestry, such as an economic recession or balance of paymeny difficulties; other measures which would be desirable may have to be deferred because of strong opposition from powerful lobbies. There is, of course, a danger that such a 'piece meal' approach may cause a lack of balance or that measures are adopted which may be prejudicial to other aspects of forest policy without this being apparent. A comprehensive approach to forest policy decisions is therefore desirable wherever it is practicable and, as will be apparent from the examples given in Chapter 8.2, countries are now increasingly adopting the practice of comprehensive periodic forest policy reviews.

Perhaps the most important policy decisions refer to priorities. More money for forestry may mean less money for agriculture or even for hospitals or schools and vice versa. Also within forestry, correct decisions on priorities ensure that best use is made of the available manpower, land and money. For example, it may be necessary to decide what priorities should be given respectively to afforestation, raising the productivity of existing forests and to soil conservation measures.

8.1.5 Implementation

As already stated, the implementation of forest policy decisions requires decisions on funding, staffing, administration, law and timing. Timing is important not only for technical reasons; sometimes people are more disposed to accept change if they are not rushed and are given proper explanations. Above all, there must be a forest authority suitably manned and equipped to implement decisions and monitor progress. The monitoring of progress not only enables achievement to be compared with what has been planned, but also indicates what further policy changes may be required.

The explanation of policy decisions is an aspect of implementation which is sometimes given too little attention. Even within forest administrations, officers are not always fully conversant with what the government has in mind. Not only they but also all others affected by forest policy, especially woodland owners, forest workers and forest industrialists are more likely to co-operate willingly in the implementation of measures if they have already been fully consulted during the process of policy formation and if, once the decisions have been taken, the decisions and the reasons for them are clearly and sympathetically presented in language which is readily understood.

Some aspects of forest policy are best dealt with in the context of a general policy on land use which determines broadly the location and type of land for afforestation, and, if applicable, for forest clearance. In the formation of such a land use policy forestry interests must obviously be given due weight.

For land use planning within the forest, a system of 'forest zoning' has been found useful in some countries. The forest authority, in consultation with the other interests concerned, prepares maps showing which objectives or functions should have priority in different forests or parts of forests: production, conservation and protection of the environment, recreation etc. This zoning may be 'indicative' i.e. indicate what is desirable or it may be 'prescriptive' i.e. prescribe what must have priority. Either way it must take into consideration the wishes and needs of the people in the area as well as the physical factors of a site.

For the successful implementation of policy the respective roles of central government, regional or local government, forest authority, woodland owners public and private must be clearly defined.

8.2 Practice

8.2.1 The need for new policies

Unprecedented changes in forest policy have been initiated in recent years or are about to be initiated in many parts of the world, and the process is still gathering momentum. The reasons are not far to seek. In the first place, as explained in Chapter 1, there is now a greater interdependence than ever before between forestry and other national policies which are evolving rapidly especially those concerned with rural development and conservation of the environment. But an even more cogent reason is that nearly everywhere the demands made on forests are increasing at an accelerating rate and it is only by new initiatives that these demands can be met without irrevocable damage being done to the forest.

We have seen that in the developing countries the main pressures arise from the population explosion which leads both to the clearance of forest for food production and an increased demand for wood and more especially for fuelwood and other wood for domestic use. Changes in policy, e.g. switching the emphasis from slow growing, high value, long rotation tree crops for export, to fast growing species to produce utility timber for local use, or the promotion of agroforestry on land where the forest has been cleared can go some way towards ameliorating the situation in the short term; but unless the problem of the population explosion is solved, changes in forest policy and practice can only delay catastrophe, they cannot avert it indefinitely. While measures to slow down population growth are not the responsibility of the forester, it is his duty to point to the consequences for forestry if population growth is not checked. Governments, in their reviews of national forest policies, and international organizations in their advice to their member countries, would do well to speak out more loudly than they have done on this delicate but fundamental issue, even at the risk of causing displeasure in some quarters. In developed countries population levels have, on the whole, become more stable, at any rate where the population density is already high. Here, increasing and, to some extent, conflicting demands are made on the forests for the:

- economic production of timber;
- conservation of the environment;
- the use of forests for recreation;
- creation of employment and improvement of living standards in rural areas.

8.2.2 The direction of change – developing countries

Some developing and recently developed countries have a long tradition of forest management. They include parts of the former British, French and Dutch colonial emprires, i.e. much of the Indian sub-continent, as well as Indonesia, Malaysia, West and East Africa. They also include a few other important countries such as Mexico. The forest histories of these countries and the evulotion of their forest policies are varied and complex, but some common trends, which are highly relevant to the policy issues of today, are not always seen in their proper perspective. Most of the foresters who established these forest services had been brought up in the Central and Western European Forestry tradition with all its virtues and faults. They regarded it as their first task to reserve appropriate areas of forest for conservation and management by the state. The emphasis was very much on conservation – protection of watersheds and stream flow etc. Management for timber production, where this was compatible with conservation, was also generally conservative; selective logging in conjunction with silvicultural measures to enrich the forest by increasing the proportion of commercially valuable species or, where there was a change to plantations, the emphasis was on long rotation species such a teak. Relatively little interest was shown for what happened outside the state forests. These other forests were mined by commercial interests, usually of the colonial power, and either converted to commercial agricultural crops or abandoned.

Little attention was paid to the needs of the local rural population for wood for fuel and other domestic purposes; this was understandable while there were no apparent shortages. Where shortages did develop, village forests were in fact established at any rate in some countries, in India already in the 19th century and elsewhere somewhat later.

In the 1950's, the emphasis of forest policy switched to large scale commercial plantations. This policy was sound as long as it did not lead to other aspects of forestry being neglected. In some countries where ruthless forest exploitation had either preceded orderly forest management or had been brought about subsequently by war or other major upheavals, the reaction was to try to make good the damage by imposing exaggerated felling restrictions or even prohibitions which usually, far from saving the forest, accelerated forest destruction for the obvious reason that the rural populations in the areas concerned depended on the forest for part or the whole of their livelihood. Mexico is a case in point. The foresters recognized that sensible management was the right answer to irrational exploitation, but successive governments supported by ignorant urban opinion considered the imposition of felling prohibitions over whole districts an easier solution. Beltran, who, while in office as Undersecretary in charge of forestry, had unsuccessfully opposed these prohibitions (called 'vedas'), summed up the consequences in his La Battalla Forestal (1964):

'Every time one of these felling prohibitions was imposed the public applauded. But the evil did not stop. Every day some new forest destruction became evident and, to combat this destruction, further prohibited zones were imposed.

But the peasant with no source of livelihood other than the forest, in spite of the most drastic measures, went on cutting trees secretly either to utilize the timber, or in the worst cases, when he was prevented from getting the timber to market, to clear a patch of land where he could sow his maize or beans'. (Translated from the original Spanish).

It was only more than 10 years later that the forest service managed to get the 'vedas' lifted and rational forest management introduced.

The broad trends of forest policy in many developing countries which have had a forest administration for a generation or more may be summed up as follows:

i The replacement of uncontrolled exploitation in selected areas (e.g. 'forest reserves') by orderly, usually rather conservative forest management with emphasis on the protective functions of the forest and the production of relatively high quality timber for industry or export.

ii During the 1950's and 1960's the emphasis tended to switch to plantations of fast growing species as a source of raw material for new major wood based industries. This was the era when industrial development was regarded as the motor for 'economic take off'.

It was only in the late 1970's that the focus of attention throughout the developing world moved towards providing the maximum immediate but sustainable benefits to the rural populations living in and near forests. 'Forests for the people', became the popular slogan. 'Forestry for the people and by the people' would be even more appropriate because no scheme can work without the active and willing participation of the intended beneficiaries. Previously, there had been limited measures with this prime objective, and of course rural populations also derived some advantage from any form of rational forest management which provided employment and the cash income that goes with it; but that was of little benefit to the far more numerous peasants on subsistence farming who need wood to cook food and to satisfy other domestic needs. This latest trend in forest policy, strongly advocated by FAO, is based, in the first place, on immediate humanitarian considerations – the principle of doing the maximum good to the maximum number, but it is now also widely believed – some would go further and say recognized – that the emphasis on commercial plantations contributed mainly to industrial and urban development, which has, if anything, made the rural poor even poorer than they were before and has failed to bring about the expected overall economic take off. More priority is therefore to be given to rural development in general and to helping the rural poor in particular. In this rural development forestry must contribute to agriculture rather than compete with it. The ambitious 'green revolution' programme of the Nigerian government illustrates this point. The main object is to raise agricultural and forestry production in the North by means of rural community developments while at the same time halting the march of the desert and thus ensuring the future production potential of the land. Another example is Mexico where the President who took office at the end of 1982 announced that rural development including forestry would be one of his top priorities.

What are the implications of the past history and the recent trends summarized above for forest policy in developing countries? Some of these implications have been discussed in section 5.7.

The most urgent and difficult problem is usually presented by the rural poor who have no land. To meet their needs, three main lines of approach have been tried or are being considered.

1 Where unused land is still available, various settlement schemes involving some combination of farming and forestry have been initiated. Experience shows that success depends very much on the partnership based on mutual trust which is built up between the settlers and the extension services that help and advise them.
2 In South Korea, landowners with more land suitable for forestry than they can themselves afforest have been encouraged to come to an arrangement with peasants without or with insufficient land, whereby the peasant does the work in return for some immediate reward and an interest in the crop. About one million ha were afforested in this way in only 5 years – an outstanding success story.
3 In Bangladesh where land is exceptionally scarce, a scheme is being considered whereby landless peasants would be encouraged to grow trees along road sides, canal banks and other strips or patches of public land. The trees would be provided free and the duty ro look after them would be accompanied by the right to use the land between the trees for pasture or to grow food or fodder crops. The controlled opening to settlement of some forest reserves which have become virtually untenable because of encroachment is also planned.

Forest policies must seek to harness the enormous latent human resource of under-employed rural populations. The impact on living conditions can be great at relatively little cost to government; moreover, such policies may contribute to general economic growth if, in addition to supplying domestic needs, a surplus is produced for sale; the beneficial effect on the environment can also be considerable. The best way to proceed must depend on local factors, and broader political issues of land tenure and land reform will have to be taken into account. There may be clashes of interest between large land owners and small holders, and even greater clashes between all who own land and landless peasants with no holding at all. On the government side, the respective roles of the forest administration and of the agricultural extension services need to be decided. As already mentioned, in some countries the forest administrations have little experience of extension work, because they have hitherto been concerned almost exclusively with the management of the state and other publicly owned forests. All this points to the need of very careful government planning.

A final word of caution: the promotion of forerstry for the people by the people must on no account lead to a neglect or, worse still, an abandonment of policies which have proved their worth in the past, and are still relevant for the future. Apart from clearly defined exceptions, existing forests, whether indigenous or planted, must continue to be managed for the production of timber and so that they may fulfil their other functions of soil and water conservation, providing habitats for wildlife etc. Additional large scale afforestation programmes by the State, other public authorities, and by industry are also as urgent as ever.

Inevitably, a major new rural forestry initiative cannot be accomodated without additional manpower and additional funds, although improved administrative procedures, aided by better communications and other new technology, may make it possible to carry out existing policies with a slightly smaller staff and thereby release some manpower for new activities.

8.2.3 The direction of change – developed countries

Forest policy, in some countries of the old world, especially in the Mediterranean region, has a very long, if little publicized, history going back 3000 years or more (Thirgood 1981). For as long as ships were built of timber, naval power depended on having access to forests and denying that access to enemies. Forests thus played a vital part in national policies. In most cases, the forests were mined but it is probable that the forests would have been destroyed much quicker if no attempts at all had been made to control exploitation. More recently, in Western Europe, the naval powers and especially the governments of Britain and France took steps to secure the management of forests to supply oak and other timbers for naval construction. Another reason why forests played such an important role in former times was that they were by far the major source of energy, not only for domestic purposes, such as cooking and heating, but also for industrial purposes, such as the smelting of metals, the production of salt and the manufacture of bricks. In some instances, the forests were simply mined and destroyed, while in others, this industrial demand was largely responsible for the introduction of a policy of sustained yield management; notable examples are the steel industry of Sweden and the salt mining industry in the Salzkammergut in Austria, both of which brought considerable areas of forest under management. Subsequently, when in some parts of Sweden the ore became exhausted, the need to find other uses for the wood led to the development of the pulp and paper industry.

In the past the demand for particular forest products has sometimes changed significantly because of technological advances, but the overall demand has continued to increase because new uses have more than compensated for those that have become obsolete. Difficulties arise when trees grown for a specific purpose are less suitable for others, e.g. coppice grown for fuel cannot yield planks and may be less suitable for pulping than specific pulpwood crops.

In much of Western and Central Europe, as the effects of uncontrolled forest exploitation became evident and governments decided to rectify the situation, the main policy aim was to organize the harvest so that there should be a sustained yield. A second objective was to improve the quantity and quality of sustainable production by better silviculture and by rehabilitating over-exploited areas. In contrast to agriculture, it was widely taken for granted that the forester should use indigenous species and keep the managed forest as close to its natural state as possible; hence the emphasis on natural regeneration, the avoidance of large clear fellings and, where possible, the favouring of mixtures rather than of monocultures. Britain and Ireland, however allowed forest destruction to proceed until 5% of the land area or less, was left under forest and then embarked after the first World War on a policy of afforestation with fast growing exotic conifers.

A little later, plantation forestry based on fast growing exotics spread to many parts of the world: South Africa, Australia, New Zealand are among the more important early examples. Many other countries, both developed and developing, followed.

The main thrust of wood production policies in most developed countries with limited forest resources has thus been to raise production quickly and economically by plantation forestry, but there are countries, especially Austria, Germany and Switzerland where production policy remains primarily geared to indigenous species, mainly because of a belief that in this way forests can best fulfil their protective and other environmental functions. These functions are of course also not ignored in other countries, where the higher production from exotic plantations has enabled some indigenous forests to be taken out of production and designated as nature reserves or wilderness areas. All these problems are obviously far less acute in sparsely populated regions with large forests, such as Canada, USA and the USSR.

8.2.4 The process of policy formation

Policy formation is a continuing process and it is not uncommon for governments to confine policy decisions at any one time to particulair issues rather than to forestry as a whole. There are good reasons for this. In the first place, politics is the art of the possible, so there is no point in holding up useful policy decisions relating to one aspect of forestry because agreement cannot be reached on others; secondly, circumstances, often outside forestry, may arise which call for immediate policy decisions. For example, a new national five year plan may have significant forest policy implications. Wise decisions are more likely to result if:

– the policy objectives have been clearly defined;
– there have been periodic reviews of the forestry sector as a whole in which the relevant facts are set out and in which the policy options and priorties are fully discussed

These two points will now be illustrated by examples

Statement of objectives. Even among the countries with a long forestry tradition there are few, if any, with a concise but comprehensive statement of forestry policy objectives, although some have attempted a short description in sufficiently general terms to accomodate changes in emphasis and interpretation. An example from the United States mentioned by Spurr (1976) is the Multiple Use-Sustained Yield Act of 1960, wich states:

> 'It is the policy of the Congress that the national forests are established and shall be administered for outdoor recreation, range, timber, watershed, and wild live and fish purposes ... The Secretary of Agriculture is authorized and directed to develop and administer the renewable surface resources of the national forests for multiple use and sustained yield of several products and services therefrom'.

Another example is provided by Great Britain where the Forestry Commission(1981) has defined its objectives as follows:

The forestry commision as the forestry authority
- To advance knowledge and understanding of forestry and trees in the countryside.
- To develop and ensure the best use of the country's forest resources; and to promote the development of the wood-using industry and its efficiency.
- To undertake research relevant to the needs of forestry.
- To combat forest and tree pests and diseases.
- To advise and assist with safety and training in forestry.
- To encourage good forestry practice in private woodlands through advice and schemes of financial assistance and by controls on felling.

The forestry commision as the forestry enterprise
- To develop its forests for the production of wood for industry by extending and improving the forest estate.
- To protect and enhance the environment.
- To provide recreational facilities.
- To stimulate and support the local economy in areas of depopulation by the development of forests, including the establishment of new plantations, and of wood-using industry .
- To foster a harmonious relationship between forestry and agriculture.
- In pursuit of the foregoing objectives, to manage the estate economically and efficiently, and to account for its activities to Ministers and Parliament.'

A more detailed and precise formulation was produced by the Commision of the

European Communities which summarized the main forest policy problems in the Community and proposed 'Principles and Objectives of Forestry Policy' which the Member States should have in common. (EEC Commision 1979). The proposed objectives and principles are reproduced in the Annex to this chapter as some of them might also be relevant elsewhere.

Short statements of policy or policy objectives as cited above are useful,mainly in order to inform the public what forestry is about in terms which they can understand, but they are usually too vague to serve as a reference point to indicate changes of policy. In drawing up statements of this kind it is not easy to strike the right balance: if the statement is too vague, it becomes meaningless; if it is too precise, it loses impact, and may require freqent change. Moreover, Governments are sometimes unenthusiastic about detailed statements of objectives because there may be aspects of policy which they may not wish to spell out; there may also be the fear that such a statement might subsequently be used as a stick with which to beat the Government if some of the objectives are not pursued with vigour,

Forest Policy Reviews. The wisdom of having periodic comprehensive reviews of forest policy is now widely recognized, but countries have differed in the methods adopted as is shown by the following examples.

One of the most far reaching reviews has recently been undertaken by the Federal Government in Canada which with its 220 million ha of productive forests accounts for almost 8% of the world total. It was found that the reserves of accessible commercial timber was far less than had hitherto been assumed, that regeneration had fallen short of expectations and that growth too was less than previously estimated. A new forest sector strategy was therefore drawn up in 1981 which is based on the concept of intensive sustainable yield management of accessible areas near industries. Apart from securing future timber supplies the new policy will cut expenditure on forest road construction and transport and will leave larger areas for nature conservation. The Provinces in whom about 90% of all forests are vested have, apparently, welcomed the Federal initative.

In the Netherlands which is one of the smallest and most densely populated countries in Europe, the Department of Economic Affairs commissioned a study to assist in the planning for forest enlargement and forest management improvement in the existing very small forest area. After considering all the relevant factors of the world forestry situation and of the country's internal circumstances the study concluded with the proposal to afforest 60,000 ha by the year 2000 and 200,000 ha by 2030.

Another comprehensive and concise review was the one commissioned by the Swiss Ministry of the Interior, which is responsible for the country's forest policy. A Commission, consisting of a group of experts, was appointed in 1971; its report entitled, 'Gesamtkonzeption fur eine Schweizerische Wald- und Holzwirtschaftspolitik' (Comprehensive concept for a Swiss Forest anf Forest Industry Policy), embraces both forestry and forest industry and concludes with a coherent set of policy proposals for each. (Eidg: Oberforstinspektorat, 1975).

The French Government adopted a broader and longer term approch, when it invited the well known philosopher, Bertrand de Jouvenel (1978) to chair a working party appointed to prepare a report 'Vers la foret du XXI siecle' (Towards the forest of the 21st century). The report concludes that the role of forests in the 21st century will continue to be of great national importance and makes a number of general proposals for the conservation, management and utilization of the forests as well as for their better intergration into national life. This broad approach has been followed up by more specific policy proposals (Duroure, 1982).

In Britain, the Centre for Agricultural Strategy of the University of Reading, under the direction of Professor J.C. Bowman and with the help of professors and lecturers from forestry and agricultural faculties at other Universities, prepared a report entitled , 'Strategy for the UK Forest Industry' (Bowman and others, 1980). In contrast to the previously cited examples, this review was not initiated by the government, but by a University. A little earlier, the Forestry Commission (1977) had reviewed one major aspect of forestry policy in its 'The Wood Production Outlook in Britain – a Review'.

In the United States, as in many other countries the development of forest policies during the 1970's centred around the conflict between timber and environment. To quote Spurr (1976):

'The controversy came sharply into focus in 1970. In June of that year, the White House released the findings and recommendations of the Task Force on Softwood Lumber and Plywood of the Cabinet Committee on Economic Policy. The primary recommendation was that the Forest Service should increase timber yields from the national forests. A goal of about 7 billion board foot (32 million cubic metres) annual increase in timber harvests by 1978 was suggested.

At the same time, clearcutting in national forests was under heavy attack from environmental groups. The Council on Environmental Quality (CEQ) had commissioned a series of reports on clearcutting in the major regions of the United States from deans of leading forestry schools. Apparently not concurring with their recommendations, the CEQ elected not to publish the reports, but rather to recommend to the President that he issue an executive order decrying clearcutting as a silvicultural practice and severely limiting its use on public forest lands.

The President was thus confronted with recommendations from his Council of Economic Advisors that logging on national forests be increased, and recommendations from his Council on Environmental Quality that could only lead to a decrease of that logging. The obvious solution was to set up a third group to adjudicate the matter and to make independent recommendations to the White House.

Thus, on 2 September 1971 President Nixon appointed the President's Advisory Panel on Timber and the Environment (PAPTE), specifically referring back to the report of the Task Force on Softwood Lumber and Plywood and requesting recommendations on the desirable level of timber harvest on federal lands and methods of accomplishing the harvest while ensuring adequate protection to the environment. Appointed to this panel were former Secretary of the Interior, Fred Seaton as Chairman; Ralph Hodges of the National Forest Products Association; Donald Zinn, who had served as President of the National Wildlife Federation; Marion Clawson from Resources for the Future; and myself. Reporting in early 1973, the Seaton Commission (or PAPTE) recommended that the levels of harvest on public forest lands could be increased, but only if funds and personnel were provided to allow the practice of much more intensive forest management on these at the same time: and that the practice of clearcutting not be regulated by law, but rather that forest managers be held responsible for harvesting timber in manners consonant with environmental standards.'

The potential influence of courts of law on forest policy is illustrated by another example from the United States described by Spurr (1976):

'National planning for uses of our public forests took on a new urgency in August of 1975 when the US 4th Circuit Court of Appeals in Richmond, Virginia upheld a district court decision that the Forest Service could not harvest trees on national forests unless they had been individually marked for cutting. The immediate effect of this ruling was to reduce national forest timber sales in a four-state area from 285 to 30 million board feet' (from 1.3 million cubic metres to 136,000 cubic metres).'

When reviewing forestry and forest industry policies it may be useful to look beyond national boundaries for advice. On questions concerning forest industry local expertise may either not be available or be too closely associated with particular commercial interests. That explains why large independent consultant firms are now employed by governments all over the world. Unfortunately, even the advice of some of these firms may not be completely objective because the views of some of the experts they employ may be coloured by links with particular industries or countries. Also, in reviewing forestry itself, some countries seek independent advice. Thus, the National Economic and Social Council of the Irish Republic commissioned Professor F.C. Convery (1979) of Duke University, USA, to undertake a broad study of Irish forest policy and to make recommendations. A way of seeking advice from abroad, which might perhaps also be worth considering elsewhere, was recently adopted by the Generalitat de Catalunya when responsibility for forestry was devolved to the region from the central government of Spain. After a preliminary review of forest policy, the local government convened a two day meeting to which senior forestry officers from five other European countries were invited. An outline of the local forestry situation and a list of specific problems had been drawn up in advance. At the meeting, the foreign participants were asked how they dealt with each of these problems in their own country and what advice, if any, they had to offer, given the situation in Catalunya. The government of the region then completed the review, taking account of the views expressed and the advice received.

Many developing countries too have embarked on comprehensive forestry policy reviews, usually with the help of the United Nations Development Programme (UNDP) and of the Food and Agriculture Organization (FAO).

A recent example is provided by Bangladesh, where several studies were carried out with such help. The studies included: the supply and consumption of wood and bamboo in Bangladesh; an inventory of the village forest resources; and the country's industrial forestry sector.

The results were then reviewed at a high level forest policy seminar in 1982, which had been organized by the Bangladesh Planning Commision in collaboration with the Ministry of Agriculture, the Forest Department, UNDP and FAO. Aid agencies with forestry interests were also invited to participatie. It may have contributed considerably to the success of the meeting that no attempt was made to reach any firm recommendations, conclusions, or decisions. The participants could thus concentrate on an objective evaluation of the options instead of being preoccupied with the defence of past policies or entrenched positions. The consensus, which emerged, did in fact point to the need for major new initiatives in forest and forest industrial policy, which could then be planned carefully by those immediately concerned.

An even more elaborate review of forest policy was undertaken by the Mexican Government during 1982-83. Separate proposals were drawn up by the Ministry of Programming and Budgets and the Ministry of Agriculture; additional proposals were prepared by various organizations and even by some individuals. FAO was then invited to send a team to Mexico to study the proposals and make recommendations.

A vital step in policy reviews, especially in developing countries, which sometimes does not receive adequate attention, is the determination of priorities and an assessment of the expected costs and benefits of particular policy proposals. The assessments must not only be in terms of money. Inputs of manpower, material and land on the one hand and benefits in terms of quality of life on the other hand must also be taken into account. Unless the means for implementation are assured, proposals must be regarded as pious hopes, rather than as firm policies.

8.2.5 Concluding remarks

We have seen that the sensible development of forest policies to meet changing needs requires a great deal of effort by many people, and that while the final responsibility for major decisions must rest with the government, wise decisions are more likely to result if the relevant facts have been assembled and analysed and if there is a continuing dialogue between the government and all the relevant non-governmental forestry interests. As it is impracticable to keep all information up to date, all the time, there is a need for periodic reviews of forest policy, but that does not necessarily mean that policy decisions on all aspects of forestry need be taken at the same time.

The role of the forest administration in policy formation is two-fold: first, to act as adviser to the government, and secondly to represent the government in its dealings with non-governmental forestry interests. Whether specifically instructed to do so or not, forest administrations should ensure that all preconditions for sensible policy decisions are met: governments have a right to expect their forest administrations to arrange for the required information to be assembled in an intelligible form and for the necessary consultative machinery to be brought into being where it does not already exist.

The role of all the non-governmental parts of the forestry sector is of course, in the first place, to safeguard their own legitimate interests, but beyond that they can make a constructive contribution to policy formation as a whole by putting at the disposal of government and of the others concerned, their knowledge and experience. To sum up, effective forest policy formation depends on:

- government interest in forestry
- an efficient forest administration;
- assembly and analysis of relevant facts;
- consultation of all concerned by government;
- collaboration between the various parts of the forestry sector;
- the integration of the forestry sector into the national economy as a whole.

References

Arnold, R.K. (1972) The Status of Forest Services within the Structure of Government Machinery – Adapting to Change 7th World Forestry Congress.

Beale, J.A. et al. (1972) Multiple Use Demands on the Forest Administrator. 7th World Forestry Congress, Buenos Aires.

Beltran, E. (1964) La Battalla Forestal. Published by author in Mexico.

Bowman, J:C: et al. (1980) Strategy for the UK Forest Industry. CAS Report 6. Centre for Agricultural Strategy, University of Reading.

CANADIAN FEDERAL GOVERNMENT (1981) Forest Sector Strategy for Canada. Ottawa, 1981.

CANADIAN FEDERAL GOVERNMENT (1982) Policy Statement – A Framework for Forest Renewal. Environment, Canada, Ottawa 1982.

COMMISSION OF THE EUROPEAN COMMUNITIES (1979) Forestry Policy in the European Community. Bulletin of the European Communities, Supplement 3/79.

Convery, F J (1779) Irish Forestry Policy National Economic and Social Council, No 46.

Donaldson, G. et al. (1978) Forerstry, a sector Policy Paper. World Bank, Washington, D.C.

Douglas, J.J. (1983) A Re-appraisal of forestry Development in Developing Countries. Martinus Nijhoff Dr W. Junk Publishers.

Duroure, R. (1982) Propositions pour une Politique Globale Forêt-Bois. Revue For. Française, Num. Sp. 1982.

FAO, (1976) A Framework for land evaluation. Soils Bulletin 32, FAO, Rome.

FAO, (1980) Towards a Forestry Strategy for Development. Committee on Forestry, Secretariat Note, April 1980.

FAO/SIDA, (1982) Report of the Seminar on Forestry Extension. FAO, Rome.

FORESTRY COMMISSION (1977) The Wood Production Outlook in Britain – a Review. Forestry Commision, Edinburgh.

FORESTRY COMMISSION (1981) The Forestry Commissions' Objectives, Policy and Procedure Paper No.1, Forestry Commission, Edinburgh.

Greeley, W.B. (1953) Forest Policy. McGraw-Hill Book Compagny, New York.

IUCN (International Union for Conservation of Nature and Natural Resources, (1980) World Conservation Strategy. IUCN-UNEP-WWF.

Johnston, D.R. (1972) The Formulation and Implementation of Forest Policy. 7th World Forestry Congress, Buenos Aires.

Jouvenel, B. (1977) Vers la Forêt du XXIe Siecle. Forêts de France, No. 215, 1978.

King, K.F.S. (FAO) (1972) A plan of action for the next 6 years. A summary of the revised FAO-study on forest policy, law and administration. FAO/7th World Forestry Congress. Proceedings pp. 1169-1188. Buenos Aires.

Laban, P. (editor) (1981) Proceedings of the workshop on land evaluation for forestry. International Institute for Land Reclamation and Improvement/ILRI, Wageningen.

Onweagba, A.E. (1981) Extension Strategy for Introducing Private Forestry in a Third World Country: an Illustration with Nigeria. In: The Indian Forester. Vol. 107 (9) pp. 563-570.

Reed, F.L.C. (1983) Forest Renewal in Canada. Commonwealth Forestry Review Vol 62 (3) No 192, Sept 1983.

Spurr, S.H. (1976) American Forest Policy in Development. University of Washington Press, Seattle, and London.

Steinlin, H. et al. (1975) Gesamtkonzeption fur eine Schweizerische Wald- und Holz-wirtschaftspolitik. Eidg. Oberforstinspektorat, 3000 Bern 14.

Thirgood, J.V. (1981) Man and the Mediterranean Forest. Academic Press, London.

Tomar, M.S. et al. (1978) Madhya Pradesh Forests and People' Demands. A situation by 2000 A.D. In The Indian Forester, vol. 104 (10) pp.661 – 674.

7th WORLD FORESTRY CONGRESS (1972) Plenary Discussion on the FAO-study on forest policy, law and administration. Proceedings, pp. 1106-1147. Buenos Aires.

Worrell. A.C. (1970) Principles of Forest Policy. McGraw-Hill, New York.

Annex

Objectives and principles of forestry policy for European community.

General principles

1. Forests should be protected and managed as a renewable resource to supply products and services which are essential to the quality of life in the European Community now and in the future. The main objectives should be:
- a sustainable increase in the economic production of timber.
- the conservation and inprovement of the environment,
- public access to forests for recreation.

Where practicable, these objectives should be pursued in conjunction with one another by multiple use management, the weight to be attached to each being varied according to ownership and the particular needs at a given place and time.

2. Forestry policy should:
- recognize the long term mature of forestry which renders sudden major changes in policy undesirable,
- take account of the distinctive characteristics and complementary roles of
– private forests,
– State forests,
– other publicly owned forests,
- seek to create conditions in which efficiently managed woodlands are economically viable.

3. Forestry policy measures should be formulated and implemented with due regard to other national and Community policies, especially those concerned with:
- land use;
- agriculture;
- wood using industries;
- regional development, including employment and standards of living,

especially in economically less-favoured regions;
- urban and rural environment.

4. Conversely, agricultural as well as other policies with possible forestry implications should pay due regard to the functions of the forest and its effective management.

5. Within the limits set by national legislation, forest owners should be free to manage their forests as they wish.

6. Forestry policy measures should be coordinated at Community level to the extent necessary to achieve common objectives.

The forest estate

1. In regions where, because of climate, topography or population density the use to which a particular piece of land is put, is of special public concern, the conversion of forest land to other land use and of other land to forestry should only be undertaken after consultation between the owner, the forest authority and other authorities concerned with land use in order to ensure a fair balance between
- forestry, farming and other land use interests,
- the interests of the owner of the land and the public interest.

2. The criteria should be clearly defined which should be taken into account when considering such changes of land use. In particular woodlands should not be regarded as land reserves but in their own right on the basis of all the products and services that forests provide to the community.

3. Measures should be taken to protect forests against serious damage by fire and other calamities and to repair the damage when a major calamity has occurred.

Wood production

1. When deciding on forest policy measures to increase wood production the expected direct return on the investment should be only one of the considerations; others should include the possible environmental benefits of forestry and the contribution which increased wood production can make to
- regional development and the living standards of rural populations, especially in less-favoured areas;
- the profitability of forest industries;
- improving the viability of forest holdings;
- cover the Community's requirements of wood.

2. The aim should be to raise the production and promote the better use of wood by measures appropriate to the particular circumstances of each country or region.

Among the measures to be considered are:
- silvicultural measures:
– accelerating the regenaration of over-mature stands;
– more general application of timely and adequate thinnings in young stands:
– choice of species and provenances suitable for the site, application of fertilizers and other measures to promote faster growth in high forest:
– conversion into productive high forest of poor quality coppice and other woodlands of low productivity:
– additional protection against fire, storm and disease:
– afforestation of bare land which is more suitable for forestry than for other purposes:
– the planting of trees outside the forest, especially of fast growing species.
- Fuller utilization of:
– trees that are harvested (branches, stumps, roots):
– wood and wood residues by the wood processing industries:
– waste paper through recycling.
- Organizational, infrastructural and institutional measures to promote efficient management, harvesting and marketing in order to reduce costs and increase revenues from wood production. Such measures could include:
– encouragement of associations of woodland owners:
– encouragement of consolidation of scattered small parcels of woodland which are in a single ownership:
– provision of roads and tracks to improve access to forests:
– market promotion and the monitoring of markets:
– the creation and development of appropriate wood processing industries within reasonable distance of the forests:
– the promotion of relevant research and development:
– the improvement of training and educational facilities.
3. Member States should draw up on a comparable basis and periodically review programmes relating to the measures listed in paragraph 2 above, giving estimates of costs and expected benefits.

Conservation of nature and protection of the human environment

1. As a minimum contribution to the conversation of nature and the protection of the human environment forests should be managed so as to
- maintain the long term fertility and productivity of the site:
- minimize the risk of causing damage elsewhere:
- take account of the landscape and wildlife.
2. Appropiate authorities should be authorized by legislation to initiate after consultation with the forest owner additional conservation measures where they are deemed necessary for specific purposes and especially for

- the protection against
- erosion by water and wind,
- desiccation and flooding,
- avalanches;
- the conservation of habitats of species of animals and plants which are in danger of extinction and whose survival is considered important.

3. As the implementation of paragraphs 1 and 2 above may add to the costs of forest management and reduce the income, the definitions and rules concerning the implementation of these paragraphs should not differ too widely between Member States and should take account of factors such as the special requirements of economically less-favoured regions.

Public access and recreation

1. Within the limits set by custom and national legislation, access on foot free of charge should be extended to as many forests as possible subject to reasonable and clearly defined exceptions in the interests of

- nature conservation, especially in areas where the survival of the forest is threatened by adverse environmental conditions:
- efficient forest management including protection against fire and damage from other causes:
- prevention of damage on adjacent areas, especially land that is farmed:
- the forest owner:
- wildlife management.

2. Where access is granted, the rights and responsibilities of the visitor, of the forest owner, of the State or other appropriate public authority should be governed by criteria which, subject to meeting specific national and local requirements, should be reasonably consistent throughout the Community.

3. The cost of the provision of recreational facilities in the forest beyond the mere granting of access on foot and from which no commercial return is to be expected should be borne by the State or other public bodies. Private forest owners should be under no obligation to provide or to let others provide such facilities in their woodlands.

Wildlife management

Subject to any Community measures which provide for more specific obligations, wildlife should be managed and controlled with the following aims in view:

- maintaining a healthy but not excessive population of as many species as are appropriate to a region and in harmony with local traditions:
- avoiding as far as possible interference with other aspects of forest manage-

ment and agriculture, especially through game damage.

Instruments of forestry policy

1. Organization
The implementation of forestry policy in each Member State should be the responsibility of a forestry authority which is effectively organized and suitably staffed for the purpose and, given the long production cycle in forestry, not too dependent on short-term fluctuations in economic and other circumstances.

2. Forestry legilisation

Member States should ensure that their forestry legislation is appropiate for the effective implementation of
- national forestry policy:
- forestry policy measures agreed at Community level.

3. Taxation and incentives

Forestry taxation and financial aids for forestry should be formulated within the general national and Community procedures for taxation and the granting of incentives, so as to provide an incentive to efficient and stable forest management, including protection against fire and other damage.

4. Research and development

The major research and development effort should be directed to solve as cost effectively as possible the most urgent problems confronting forest management by
- careful choice of research priorities;
- cooperation and coordination at both national and Community levels, where this is likely to result in a worthwhile economy of effort;
- the promotion at Community level of selected research projects of particular importance and beyond the capacity of individual national effort.

5. Education and training

Member States should ensure that adequate education and training facilities in forestry, including refresher courses, are available either nationally or by arrangement with institutions elsewhere. The facilities should
- cover all aspects of forestry;

- seek to achieve a reasonable balance between supply of and demand for personnel;
- meet the requirements of forest owners and other employers as well as of all categories of employees;
- bring about a gradual approximation of qualifications and standards throughout the Community as a means of facilitating the mutual recognition of qualifications and the free movement of personnel at all levels in accordance with the accepted social policy of the Community.

6. Information

Member States should exploit and develop the necessary statistics on forests on the basis of criteria and definitions common to all the Member States in order to:
– ensure that national statistics are comparable and
– enable statistics useful at Community level to be grouped together.
Exchanges of information other than statistics within the Community should also be intensified.

7. Consultation

Measures should be taken, where they do not already exist, to provide for frequent consultations at national level between the forestry authority and the organizations representing:
– owners of private and public forests,
– employees,
– the primary processing industries,
– the timber trade,
– those concerned with conserving nature and the landscape.
Consultations at Community level between these various interest groups should also be encouraged.

8. Public relations

Steps should be taken to give the public as a whole, and young people in particular, a better understanding of all aspects of forests.